Acta Anatomica

International Journal of Anatomy, Embryology and Cell Biology

Founded 1945 by R. Chambers, G. Glimstedt, T. Peterfi and G. Wolf-Heidegger
Continued 1962–1974 by E.A. Boyden, 1955–1980 by A. Delmas,
1972–1980 by F. Walberg, 1945–1980 by G. Wolf-Heidegger,
1981–1988 by R. O'Rahilly, Davis, Calif., 1989–1990 by G.E. Goslow, Jr., Providence, R.I.,
1981–1992 by W. Lierse, Hamburg

KARGER

Printed in Switzerland
on acid-free paper by
Reinhardt Druck, Basel

Appears bimonthly:
1 volume per year
(6 issues)

Guidelines for Authors

Acta Anatomica

Aims and Scope

Acta Anatomica is an international forum for communication among scientists interested in morphology at all levels of organization, from subcellular to macroscopy, and in morphogenesis, with emphasis on humans and higher vertebrates. The journal offers rapid exchange of novel and significant information gained from experimental and theoretical investigations on all aspects that are pertinent to the formation and transformation of biological structure.

Topics of special interest include

Vertebrate embryology (including developmental and reproductive biology), cell biology, neuroanatomy, functional and clinical anatomy (including application of modern techniques of imaging). Papers highlighting the links between structural and functional aspects are preferred.

Types of Papers

The journal publishes: (1) Original papers on experimental research. (2) Theoretical contributions. These will appear in the section 'Hypotheses and Concepts' and may include provocative hypotheses, provided that they are innovative, well founded and show promise for new avenues of research. (3) Short reviews/perspectives. While (1) and (2) are submitted contributions, reviews/perspectives (3) are invited. Suggestions concerning reviews on current topics of high interest are welcome.
All articles submitted must be original. All manuscripts must be approved by all authors, and this has to be certified by the first author with his signature. Manuscripts may not have been published previously either in whole or in part, except in abstract form, and may not be under consideration by any other journal. Reported studies using animal and human subjects must have been approved by the appropriate institutional review boards.

Submission

Manuscripts from North and South America should be sent to:

Prof. A.W. English, PhD
Emory University School of Medicine
Dept. of Anatomy and Cell Biology
Atlanta, GA 30322 (USA)
Telefax +1 404 727-3677
E-Mail: acta@cellbio.emory.edu

Manuscripts from Japan and South-East Asia to:

Prof. Dr. Osamu Ohtani
Department of Anatomy II
Toyama Medical and Pharmaceutical University
2630 Sugitani
Toyama, 930-01 (Japan)
Telefax +81 764 34 5010

Manuscripts from all other countries to:

Prof. Dr. med. Dr. rer. nat. H.-W. Denker
c/o S. Karger AG
Editorial Dept. 'Acta Anatomica'
P.O. Box
CH-4009 Basel (Switzerland)
Telefax +41 61 306 14 34

All manuscripts are subject to peer review. Submission of an article for publication implies the transfer of the copyright from the author to the publisher upon acceptance. Accepted papers become the permanent property of Acta Anatomica and may not be reproduced by any means, in whole or in part, without the written consent of the publisher. It is the author's responsibility to obtain permission to reproduce illustrations, tables, etc. from other publications.

Format and Arrangement

Manuscripts should be written in English and be as concise as possible. They should be submitted in quadruplicate (with four sets of original illustrations), typewritten double-spaced on one side of the paper, with a wide margin. Lines should be numbered on the margin.

Title page: Title (main title underlined), the authors' names, and the institute where the work was conducted, a short title for use as running head, the full address of the author to whom correspondence should be sent. Please also supply phone and fax numbers, as well as your e-mail address. If different, indicate in addition a full address for revisions and proof reading separately.

Key words: For indexing purposes provide a list of up to 5 key words.

Abstract: Each paper needs an abstract of up to about 20 lines summarizing the essential new informations communicated.

Subheadings: Introduction; Materials and Methods; Results; Discussion.

Small type: Paragraphs which can or must be set in smaller type (methods, etc.) should be indicated with a 'p' (petit) in the margin on the left-hand side.

Footnotes: Abbreviations used in the paper will be printed as a footnote on the first page. Please list them on a separate page of the manuscript. Otherwise, footnotes should be avoided. When essential, they are numbered consecutively and typed at the foot of the appropriate page.

Nomenclature: The nomenclature used should conform, wherever possible, to the current edition of the Nomina Anatomica, Nomina Anatomica Veterinaria, or to the Nomina Anatomica Avium.

Tables and illustrations: Tables and illustrations (both numbered in Arabic numerals) should be prepared on separate sheets. Tables require a heading and figures a legend, also prepared on a separate sheet. For the reproduction of illustrations, only good drawings and original photographs can be accepted; negatives or photocopies cannot be used. Due to technical reasons, figures with a screen background should not be submitted. When possible, group several illustrations on one block for reproduction (max. size 181 × 223 mm) or provide crop marks. On the back of each illustration, indicate its number, the author's name, and 'top' with a soft pencil.

Color illustrations: During the review process, the decision will be made as to which color illustrations of a manuscript can be reproduced free of charge. Authors will be informed of this at the time of acceptance of their paper. Otherwise, these illustrations will be reproduced at the author's expense. Up to 6 color illustrations per page can be integrated into the text at the reduced price of CHF 660.– / USD 545.00 per page.

References: References in the text should be made by author's name followed by the year of publication, arranged chronologically, then alphabetically. When there are more than two authors, use the first author's name and et al. When references are made to more than one paper by the same author, published in the same year, designate them as a, b, c, etc. The list of references should include only those publications which are cited in the text. Material submitted for publication but not yet accepted may be cited in the text as 'unpublished data' but must not be included in the reference list. In the list of references, arrange references alphabetically listing all authors, then year of publication. Abbreviate journal names according to Index Medicus.

Examples
(a) Papers published in periodicals: Sun, J., H. Koto, K.F. Chung (1997) Interaction of ozone and allergen challenges on bronchial responsiveness and inflammation in sensitised guinea pigs. Int Arch Allergy Immunol *112:* 191–195.
Hadler, W.A., S.R. Silveira (1981) Técnica histoquímica que permite discriminar o colesterol de substâncias afins e de seus ésteres. Cienc Cult 33(suppl): 586–587.
(b) Monographs: Matthews DE, Farewell VT: Using and Understanding Medical Statistics, ed 3, revised. Basel, Karger, 1996.
(c) Edited books: Parren, P.W.H.I., D.R. Burton (1997) Antibodies against HIV-1 from phage display libraries: Mapping of an immune response and progress towards antiviral immunotherapy; in Capra JD (ed): Antibody Engineering. Chem Immunol. Basel, Karger, vol 65, pp 18–56.

Further details: Authors may consult our leaflet 'Rules for the Preparation of Manuscripts' available on request from the Publisher, S. Karger AG, P.O. Box, CH-4009 Basel (Switzerland), and the 'Guide to Writing Articles in English' by R. O'Rahilly (Acta anat. *131,* 1–2, 1988).

Cover illustrations

The Journal Acta Anatomica will publish cover illustrations relevant to any area of modern anatomical research. Cover illustrations will be chosen primarily on the basis of aesthetic appeal and not necessarily from articles in that issue of the journal. Authors or readers are encouraged to submit illustrations suitable for the cover. Only black-and-white illustrations can be accepted. Please submit two copies (max. 13 × 16 cm) to the appropriate editors. Submissions should include a brief caption on the back of the illustration.

Galley proofs

Unless indicated otherwise, galley proofs are sent to the first-named author and should be returned with the least possible delay. Alterations made in proofs, other than the correction of printer's errors, are charged to the author.

Reprints

Order forms and a price list are sent with the proofs. Orders submitted after the issue is printed are subject to considerably higher prices. Allow five weeks from date of publication for delivery of reprints.

KARGER

Fax + 41 61 306 12 34
E-Mail karger@karger.ch
www.karger.com

The Guidelines for Authors are available at:
www.karger.com/journals/aan/aaninst.htm

General Information

Acta Anatomica

Publication data: 'Acta Anatomica' is published 12 times annually. Volumes 158–159 with 4 issues each appears in 1997.

Subscription rates: Subscriptions run for a full calendar year. Prices are given per volume.
Personal subscription:
Subscription + postage and handling
CHF 359.50 + CHF 15.70 Europe
CHF 359.50 + CHF 23.70 Overseas
USD 299.50 + USD 15.70
(Must be in the name of, billed to, and paid by an individual. Order must be marked 'personal subscription'.)
Institutional subscription:
Subscription + postage and handling
CHF 719.– + CHF 19.60 Europe
CHF 719.– + CHF 29.60 Overseas
USD 599.00 + USD 19.60
Airmail postage: CHF 24.–, USD 20.00.
(Extra per volume.)

Microforms: Please direct all enquiries and orders to: University Microfilms Inc.
300 North Zeeb Road, P.O. Box 1346
Ann Arbor, MI 48106-1346 (USA)

Subscription orders: Orders can be placed at agencies, bookstores, directly with the Publisher
S. Karger AG, P.O. Box,
CH–4009 Basel (Switzerland)
or with any of the following distributors:
Deutschland: S. Karger GmbH, Postfach
D–79095 Freiburg
France: Librairie Luginbühl
36, bd de Latour-Maubourg, F–75007 Paris
Great Britain, Ireland: S. Karger Publishers
58 Grove Hill Road
Tunbridge Wells, Kent TN1 1SP
USA: S. Karger Publishers, Inc.
26 West Avon Road, P.O. Box 529
Farmington, CT 06085
India, Bangladesh, Sri Lanka:
S. Karger India Subscription Agency
B–5/132 Safdarjung Enclave
New Delhi-110029
Japan: Katakura Libri, Inc.
Take 1 Bldg, 2nd Floor, 4-5-7 Yushima
Bunkyo-ku, Tokyo 113

Change of address: Both old and new address should be sent to the subscription source.

Single issues and back volumes: Information can be obtained through the Publisher.

Advertising: Correspondence and rate requests should be addressed to the Publisher.

Bibliographic indices: This journal is regularly listed in bibliographic services, including *Current Contents®*.

Copying: This journal has been registered with the Copyright Clearance Center (CCC), as indicated by the code appearing on the first page of each article.
For readers in the US, this code signals consent for copying of articles for personal or internal use, or for the personal or internal use of specific clients, provided that the stated fee is paid per copy directly to
Copyright Clearance Center Inc.
222 Rosewood Drive
Danvers, MA 01923 (USA)
A copy of the first page of the article must accompany payment. Consent does not extend to copying for general distribution, for promotion, for creating new works, or for resale. In these cases, specific written permission must be obtained from the copyright owner,
S. Karger AG,
P.O. Box, CH–4009 Basel (Switzerland).

KARGER

Fax+ 41 61 306 12 34
E-Mail karger@karger.ch
www.karger.com

The Journal Home Page is available at:
www.karger.com/journals/aan

Acta Anatomica

Functional Implications of the Design of Skeletal Muscles

Guest Editors
Roland R. Roy, Los Angeles, Calif.
Akihiko Ishihara, Kyoto

49 figures, 16 tables, 1997

KARGER

Basel · Freiburg · Paris · London · New York ·
New Delhi · Bangkok · Singapore · Tokyo · Sydney

KARGER

S. Karger
Medical and Scientific Publishers
Basel · Freiburg · Paris · London
New York · New Delhi · Bangkok
Singapore · Tokyo · Sydney

Drug Dosage
The authors and the publisher have exerted every effort to ensure that drug selection and dosage set forth in this text are in accord with current recommendations and practice at the time of publication. However, in view of ongoing research, changes in government regulations, and the constant flow of information relating to drug therapy and drug reactions, the reader is urged to check the package insert for each drug for any change in indications and dosage and for added warnings and precautions. This is particularly important when the recommended agent is a new and/or infrequently employed drug.

P.O. Box, CH–4009 Basel (Switzerland)
Printed in Switzerland on acid-free paper by
Reinhardt Druck, Basel
ISBN 3–8055–6682–4

Contents Vol. 159, No. 2–3, 1997

Acta Anatomica

KARGER

Fax+ 41 61 306 12 34
E-Mail karger@karger.ch
www.karger.com

Access to full text and tables of contents,
including tentative ones for forthcoming issues:
www.karger.com/journals/aan/aancont.htm

Acta Anatomica

Acta Anat 1997;159:75–77

R.R. Roy [a]
A. Ishihara [b]

[a] Brain Research Institute, UCLA School of Medicine, Center for the Health Sciences, Los Angeles, Calif., USA;
[b] Laboratory of Neurochemistry, Faculty of Integrated Human Studies, Kyoto University, Kyoto, Japan

Overview: Functional Implications of the Design of Skeletal Muscles

Two of the features of a skeletal muscle that have been shown to be important in determining its function are its architectural design and fiber-type composition. For example, the two most reliable and reproducible indices of the functional capacities of a skeletal muscle are its maximum force production and its maximum rate of shortening determined under maximal stimulation conditions. The architectural design of the skeletal muscle has a significant impact on each of these functional properties. First, consider two muscles having similar masses and biochemical profiles, but different architectural features. One muscle has relatively long fibers and a minimal angle of fiber pinnation; the other has relatively short fibers and the fibers are highly pinnated. Under maximal stimulation conditions and identical biomechanical constraints, the relative proportion of half sarcomeres arranged either in series or in parallel will dictate the relative priority for force and displacement in each muscle. Thus, the design of the first muscle would optimize displacement (a relatively high percentage of half sarcomeres arranged in series), whereas the design of the second muscle would optimize force production (a relatively high percentage of half sarcomeres arranged in parallel). This is just one simple example of how the architectural features of a muscle can significantly impact its function. A similar argument can be used to describe how the fiber type composition of a muscle, which reflects its biochemical potential, influences its function. Consider two muscles having similar architectural properties, but different fiber-type compositions. The muscle having a higher proportion of fast fibers will have a faster shortening velocity than the muscle having a higher proportion of slow fibers.

In the past 15 years, the combination of refined morphological techniques and renewed interest in morphological-functional interrelationships in mammalian skeletal muscle have resulted in dramatic advances in the identification and understanding of the critical motor control issues at the muscle, motor unit and muscle fiber levels [see Josephson, 1975; Partridge and Benton, 1981; Sacks and Roy, 1982; English and Letbetter, 1982; Stein, 1982; Edgerton, 1983, 1986, 1987; Loeb et al., 1987; English and Weeks, 1987; Windhorst et al., 1989; Ounjian et al., 1991; Roy et al., 1995 for a discussion of some of these issues]. For example, an excellent review of the earlier literature on the theoretical effects of muscle architecture-function interrrelationships was published by Gans and Bock in 1965. Since then, several reviews in the general area of muscle architecture-function relationships have been published [Close, 1972; Gans, 1982; Edgerton et al., 1983; Edgerton et al., 1986; Otten, 1988; Peters, 1989; Roy and Edgerton, 1992; Gans and Gaunt, 1992; Trotter, 1993; Fukunaga et al., 1997]. Similarly, the functional correlates of the different fiber types have been studied at the whole muscle [Close, 1972; Burkholder et al., 1994], motor unit [Burke and Edgerton, 1975; Burke, 1981; Bodine et al., 1987] and single fiber levels [Peter et al., 1972; Bottinelli et al., 1991; Talmadge et al., 1993].

In this issue, we present a series of 10 papers addressing structure-function interrelationships, particularly at the whole muscle level. Fukunaga and co-workers used ultrasonography to visualize the changes in the fascicles and aponeuroses of the human vastus lateralis muscle during ramp isometric contractions. They found that pinnation angle and fascicle lengths changed significantly during these isometric contractions, although the muscle-tendon unit was maintained at a constant length. These data are thought to reflect the compliance within the muscle-tendon

KARGER
Fax+41 61 306 12 34
E-Mail karger@karger.ch
www.karger.com

0001–5180/97/1593–0075$15.00/0

This article is also accessible online at:
http://BioMedNet.com/karger

Roland R. Roy, PhD
Brain Research Institute, UCLA School of Medicine, Center for the Health Sciences
10833 Le Conte Avenue, Los Angeles, CA 90095-1761 (USA)
Tel. +1 310 825 2193, Fax +1 310 206 9184
E-Mail rrr@ucla.edu

complex and indicate that these architectural parameters should be considered in designing models of muscle mechanical properties. Lieber determined the architectural properties of mouse plantarflexor and dorsiflexor muscles and presented a dynamic model to study the coordination between mean fiber lengths and moment arms during normal movements. The data indicate that, in general these antagonistic groups are architecturally designed to optimize the appropriate forces and velocities required to perform normal locomotion. Ohira and co-workers immobilized the soleus muscle in either a shortened or lengthened state in hindlimb unloaded rats and reported adaptations in the fiber type and metabolic properties that were closely associated with chronically-imposed changes in muscle length. These data provide evidence that the chronic loading characteristics on a muscle can significantly affect the amount and type of protein metabolism in skeletal muscles.

Hijikata and Ishikawa review the literature on the presence of serially-linked muscle fibers and of nonspanning tapering muscle fibers that terminate intrafascicularly in a number of mammalian muscles. The authors discuss the potential modes of transmission of forces in these muscle fibers and the potential role of these fiber arrangements in myopathic conditions. These data have important implications for motor control strategies involving muscles having these types of fiber arrangements. Tamaki's group show an increase in the number of complex branched fibers, 10 or more muscle fibers fused together and having a large number of daughter branches, in the rat plantaris muscle following degeneration-regeneration induced by repeated intramuscular marcaine injections. The presence of a large number of complex branched fibers was associated with a 'slowing' of the twitch properties, without any change in the percent of slow fibers. Thus, it appears that complex branched fibers impact the mechanical properties of the normally fast plantaris via a mechanism independent of fiber type composition. Eldred and co-workers determined the interrelationships among the architectural design, the fiber type composition and the muscle spindle density of the cat capsularis, a small hip muscle located close to the joint capsule. The potential influence of sensory input (particularly spindle discharge) from the regions of high density of muscle spindles in determining the location of the relatively few fast fibers in this predominantly slow muscle was explored. Both the superficial to deep and the transverse gradients in fiber type distributions were consistent with the presumed gradients of spindle discharge based on anatomical considerations.

Hermanson describes the gross morphology and fiber type distribution in the extensor carpi radialis muscle of the horse. A large fast to slow gradient in the fiber type distribution from the proximal to the distal end of the muscle, a form of compartmentalization, is reported. In addition, a unique arrangement between the slow distal compartment of the extensor carpi radialis and a predominantly slow distal slip of the biceps brachii (the lacertus fibrosus) exists, which is hypothesized to serve as a passive stay-apparatus to effect long-term standing. A similar type of struture-function relationship is described by Roy's group for a number of muscles located around the cat hip, called hip 'cuff' muscles by these authors. Compared to other hindlimb muscles studied, all of these deeply located muscles had short fibers, small angles of pinnation and, at least in some regions, high proportions of slow fibers. These fiber-type distributions and architectural features are consistent with these muscles serving as stabilizers of the hip. Chleboun and co-workers describe the architectural and fiber-type distributions of the four bellies of the mouse extensor digitorum muscle. Although the fiber type composition was similar across the four bellies, the muscle belly of digit 5 had the largest mass and physiological cross-sectional area and the longest fibers. These data suggest that digit 5 would produce the highest tensions and have the fastest maximal rate of shortening during in vivo movements.

Ishihara's group determined the type, size, and succinate dehydrogenase activity, an oxidative enzyme marker, of the fibers in two perineal muscles, i.e. the levator ani and bulbocavernosus. In addition, using retrograde labeling, the motoneurons innervating these muscles were identified and their size and succinate dehydrogenase activities determined. Relative to other motor pools, both the mean motoneuronal soma size and muscle fiber cross-sectional areas were relatively small. However, the mean succinate dehydrogenase activities were mismatched, being relatively high in the motoneurons and relatively low in the muscle fibers that they innervate. These data indicate that the metabolic potentials of the components of a neuromuscular unit need not be coordinated, even in normal sedentary animals.

In effect, the papers included in this issue represent a cross section of some critical structure-function issues being addressed at the whole skeletal muscle level. We feel that the results discussed in this series of papers include important implications which should be considered in the design of future studies related to skeletal muscle function and motor control. The editors would sincerely like to thank the contributors and reviewers for their participation in this endeavor. We also thank Dr. A.W. English, Editor-in-Chief for *Acta Anatomica,* for his continued support, advice and patience and Miss Jung Kim at UCLA for her expert technical assistance.

References

Bodine, S.C., R.R. Roy, E. Eldred, V.R. Edgerton (1987) Maximal force as a function of anatomical features of motor units in the cat tibialis anterior. J Neurophysiol *57:* 1730–1745.

Bottinelli, R., S. Schiaffino, C. Reggiani (1991) Force-velocity relationship and myosin heavy chain isoform compositions of skinned fibres from rat skeletal muscle. J Physiol (Lond) *437:* 655–672.

Burke, R.E. (1981) Motor units: Anatomy, physiology, and functional organization; in Brookhart JM, VB Mountcastle (eds): Handbook of Physiology, sect 1, The Nervous System, vol 2, Motor Control, part 2, chap 10. Maryland, American Physiological Society, pp 345–422.

Burke, R.E., V.R. Edgerton (1975) Motor unit properties and selective involvement in movement; in Wilmore JH, J. Keogh (eds): Exercise Sport Sciences Reviews. New York, Academic Press, vol 3, pp 31–81.

Burkholder, T.J., B. Fingado, S. Baron, R.L. Lieber (1994) Relationship between muscle fiber types and sizes and muscle architectural properties in the mouse hindlimb. J Morphol *221:* 177–190.

Close, R.I. (1972) Dynamic properties of mammalian skeletal muscle. Physiol Rev *52:* 129–197.

Edgerton, V.R., S.C. Bodine, R.R. Roy (1987) Muscle architecture and performance: Stress and strain relationships in a muscle with two compartments arranged in series; in Marconnet P, PV Komi (eds): Medicine and Sport Science: Muscular Function in Exercise and Training. Basel, Karger, vol 26, pp 12–23.

Edgerton, V.R., R.R. Roy, S.C. Bodine, R.D. Sacks (1983) The matching of neuronal and muscular physiology; in Borer KT, DW Edington, TP White (eds): Frontiers of Exercise Biology. Chicago, Human Kinetics Publishers, pp 51–70.

Edgerton V.R., R.R. Roy, R.J. Gregor, S. Rugg (1986) Morphological basis of skeletal muscle power output; in Jones NL, N McCartney, AJ McComas (eds): Human Muscle Power. Chicago, Human Kinetics Publishers, pp 43–64.

English, A.W., W.D. Letbetter (1982) Anatomy and innervation patterns of cat lateral gastrocnemius and plantaris muscles. Am J Anat *164:* 67–77.

English, A.W., O.I. Weeks (1987) An anatomical and functional analysis of cat biceps femoris and semitendinosus muscles. J Morphol *191:* 161–175.

Fukunaga, T., Y. Kawakami, S. Kuno, K. Funato, S. Fukashiro (1997) Muscle architecture and function in humans. J Biomech *30:* 457–463.

Gans, C. (1982) Fiber architecture and muscle function. Ex Sport Sci Rev *10:* 160–207.

Gans, C., W.J. Bock (1965) The functional significance of muscle architecture: A theoretical analysis. Ergeb Anat Entwicklungsgesch *38:* 115–142.

Gans, C., A.S. Gaunt (1992) Muscle architecture and control commands. Brain Behav Evol *40:* 70–81.

Josephson, R.K. (1975) Extensive and intensive factors determining the performance of striated muscle. J Exp Zool *194:* 135–154.

Loeb, G.E., C.A. Pratt, C.M. Chanaud, F.J.R. Richmond (1987) Distribution and innervation of short, interdigitated muscle fibers in parallel-fibered muscles of the cat hindlimb. J Morphol *191:* 1–15.

Otten, E. (1988) Concepts and models of functional architecture in skeletal muscle. Ex Sport Sci Rev *16:* 89–137.

Ounjian, M., R.R. Roy, E. Eldred, A. Garfinkel, J.R. Payne, A. Armstrong, A.W. Toga, V.R. Edgerton (1991) Physiological and developmental implications of motor unit anatomy. J Neurobiol *22:* 547–559.

Partridge, L.D., L.A. Benton (1981) Muscle, the motor; in Brookhart JM, VB Mountcastle, VB Brooks, SR Geiger (eds): Handbook of Physiology, sect 1, The Nervous System. Bethesda, American Physiological Society, pp 43–106.

Peter, J.B., R.J. Barnard, V.R. Edgerton, C.A. Gillespie, K.E. Stempel (1972) Metabolic profiles of three fiber types of skeletal muscle in guinea pigs and rabbits. Biochemistry *11:* 2627–2634.

Peters, S.E. (1989) Structure and function in vertebrate skeletal muscle. Am Zool *29:* 221–234.

Roy, R.R., V.R. Edgerton (1992) Skeletal muscle architecture and performance; in Komi PV (ed): Strength and Power in Sport: Encyclopedia of Sports Medicine. Oxford, Blackwell Scientific Publications, vol 3, pp 115–129.

Roy, R.R., A. Garfinkel, M. Ounjian, J. Payne, A. Hirahara, E. Hsu, V.R. Edgerton (1995) Three-dimensional structure of cat tibialis anterior motor units. Muscle Nerve *18:* 1187–1195.

Sacks, R.D., R.R. Roy (1982) Architecture of the hindlimb muscles of cats: Functional significance. J Morphol *173:* 185–195.

Stein, R.B. (1982) What muscle variable (s) does the nervous system control in limb movements? Behav Brain Sci *5:* 535–577.

Talmadge, R.J., R.R. Roy, R.R., V.R. Edgerton (1993) Muscle fiber type and function. Curr Opin Rheumatol *5:* 695–705.

Trotter, J.A. (1993) Functional morphology of force transmission in skeletal muscle. Acta Anat *146:* 205–222.

Windhorst, U., T.M. Hamm, D.G. Stuart (1989) On the function of muscle and reflex partitioning. Behav Brain Sci *12:* 629–681.

Acta Anatomica

Acta Anat 1997;159:78–83

Y. Ichinose
Y. Kawakami
M. Ito
T. Fukunaga

Department of Life Sciences (Sports Sciences), University of Tokyo, Japan

Estimation of Active Force-Length Characteristics of Human Vastus lateralis Muscle

Key Words
Ultrasound
Force-length relationship
Isometric contraction
Tendon compliance

Abstract
The length and angles of fascicles were determined for the vastus lateralis muscle (VL) using ultrasonography in 6 subjects performing ramp isometric knee extension. The subject increased torque from zero (relax) to maximum (MVC) with the knee positioned every 15°, from 10° to 100° flexion (0° = full extension). As the knee was positioned closer to extension, fascicle length was shorter [116 ± 4.7 (mean ± SEM) mm at 100° vs. 88 ± 4.1 mm at 10° (relax)]. The fascicle length of the VL decreased with increasing torque at each knee position [116 ± 4.7 (relax) to 92 ± 4.3 mm (MVC) at 100°]. On the other hand, fascicle angles increased with an increase in torque. These changes reflected the compliance of the muscle-tendon complex which increased as the knee reached a straight position. The estimated muscle force of the VL was maximal (2,052 ± 125 N) for a fascicle length of 78 ± 2.7 mm (i.e. optimum length) with the knee positioned at 70° of flexion. The relationship between muscle force and fascicle length indicated that the VL uses the ascending (knee <70°), plateau (70°), and descending regions (>70°) of the force-length curve.

Introduction

The length of a muscle fiber depends on the number of sarcomeres in series with the fiber. The relationship between the length and tension of a muscle fiber, therefore, reflects that of the sarcomeres, generating maximal tension at optimal sarcomere length [Woittiez et al., 1983; Huijing et al., 1989; Lieber and Brown, 1992a].

Muscle fibers are packed in a fascicle which is connected to tendinous tissue (aponeurosis and/or tendon). These tissues comprise the muscle-tendon complex (MTC), with contractile (i.e. muscle fibers, CC) and elastic (i.e. tendinous tissue, EC) components. The length changes of the MTC and muscle fibers are not equivalent due to the fact that the muscle fibers are arranged diagonally to the aponeurosis, and that the tendinous tissue is elongated during contraction [Muhl, 1982]. Therefore, muscle fiber length cannot be easily determined from the joint angle.

Previous studies have obtained fiber length and pennation angles of human muscles from cadavers [An et al., 1981; Wickiewicz et al., 1983; Friedrich and Brand, 1990; Lieber and Brown, 1992b]. Recently, a technique has been developed to determine fascicle length and fascicle angles in humans in vivo using ultrasonography [Rutherford and

KARGER
Fax + 41 61 306 12 34
E-Mail karger@karger.ch
www.karger.com

0001–5180/97/1593–0078$15.00/0

This article is also accessible online at:
http://BioMedNet.com/karger

Tetsuo Fukunaga, PhD
Department of Life Sciences (Sport Sciences), University of Tokyo
Komaba 3-8-1, Meguro, Tokyo 153 (Japan)
Tel./Fax +81-3-5454-6869
E-Mail fukunaga@idaten.c.u-tokyo.ac.jp

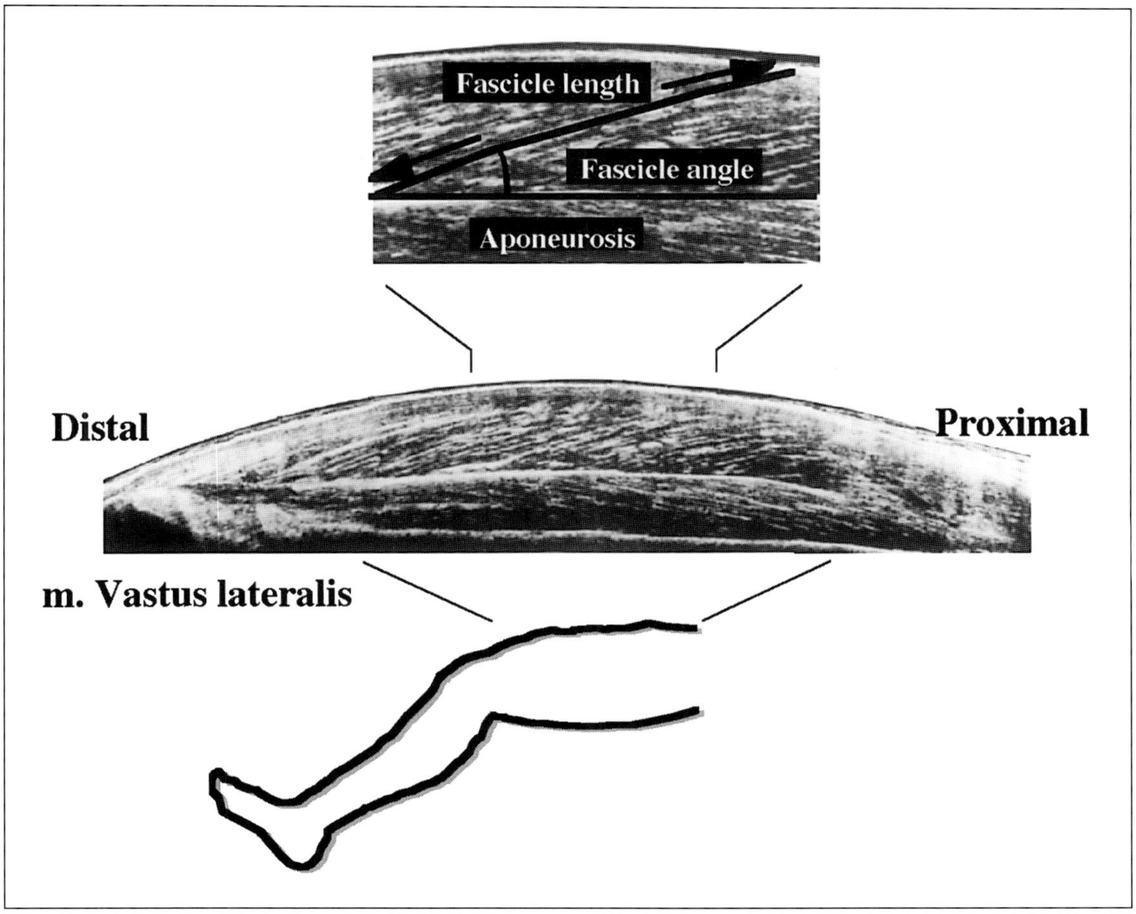

Fig. 1. Transverse ultrasonic image of the vastus lateralis muscle.

Jones, 1992; Kawakami et al., 1993; Herbert and Gandevia, 1995; Narici et al., 1996; Fukunaga et al., 1996, 1997]. Our preceding work [Fukunaga et al., 1997] demonstrated that the relationships between fascicle lengths and joint angles differ in relaxing and isometrically contracting muscles, possibly due to the compliance of tendinous tissue. In fact, shortening of muscle fibers during isometric contraction at the cost of taking up tendon compliance has been demonstrated in animal studies [Griffiths, 1987]. It is expected, therefore, that the relationship between force and length of contracting human muscles would differ from those reported previously based on cadaveric materials [Cutts, 1988; Herzog and ter Keurs, 1988]. The purpose of the present study was to measure fascicle length and fascicle angle in actively contracting muscle to calculate the active force-length characteristics of a human muscle in vivo, and to investigate the influence of the compliance of the MTC.

Methods

Subjects

Six males (23.7 ± 0.8 years; 168.5 ± 5.2 cm; 65.8 ± 5.7 kg; mean ± SD) volunteered as subjects. This study was approved by the office of the Department of Sports Sciences, University of Tokyo, and complied with their requirements for human experimentation. The purpose and possible risks associated with these measurements were explained to the subjects before obtaining their consent to participate in the study.

Protocol

The subject was seated on a test bench of an electric dynamometer (MYORET, RZ-450, ASICS, JPN) with the hip joint at 80° (full extension; 0°). The axis of the lever arm of the dynamometer was visually aligned with the center of rotation of the knee joint, and the lever arm was fixed by a cuff to the lower leg just proximal to the lateral malleolus. The knee joint angle was set every 15° from 100° (flexed position) to 10° (extended position).

Following a warm-up session with submaximal contractions, the subject was instructed to develop a ramp isometric contraction, exert-

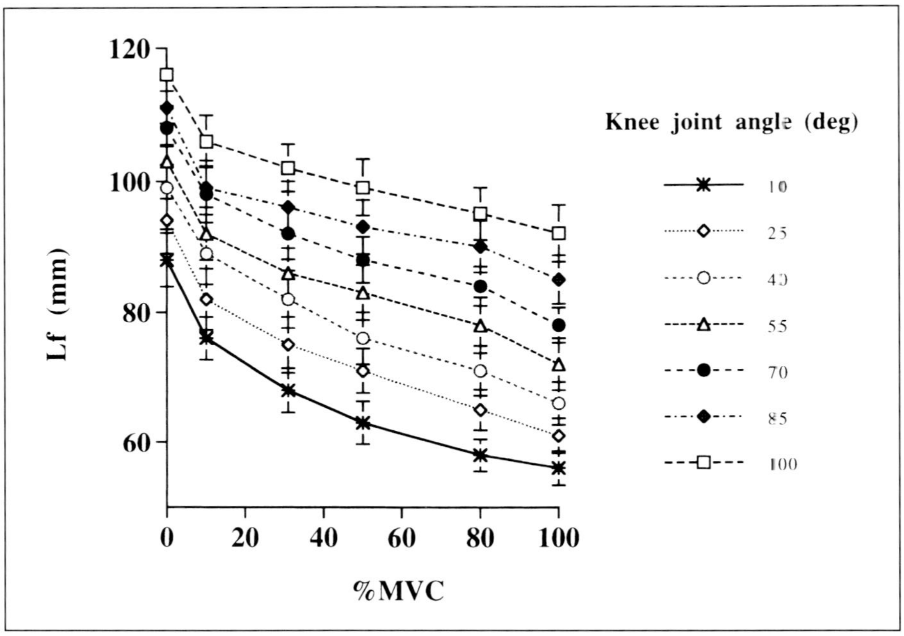

Fig. 2. The relationship between fascicle length (Lf) and knee extension torque (%MVC) at different knee joint angles. Symbols represent mean and SEM.

ing torque from zero (relax) to voluntary maximum (MVC) in 5 s at each joint angle. This test was performed at least twice. A third trial was performed if the subject or experimenters judged that the contraction was not maximal. The mass of the lower limb and the passive torque were subtracted from the exerted torque. The highest peak torque among the trials was adopted. The order of testing was randomized for knee positions, and the subject had a 2-min rest between tests to minimize fatigue. Torque output was A/D converted at a sampling rate of 100 Hz and analyzed by a personal computer.

Ultrasonic Measurement

The muscle tested was the vastus lateralis (VL). This is a unipennate muscle having a flat muscle belly and has the largest physiological cross-sectional area (PCSA) and the greatest force production among the quadriceps femoris muscles [Lieb and Perry, 1968; Narici et al., 1989; Akima et al., 1995].

An ultrasonic apparatus (SSD-2000, Aloka, JPN) with a mechanical linear scanner (scanning frequency; 7.5 MHz) was used to visualize the fascicles and aponeuroses of the VL [Fukunaga et al., 1997]. The measurement site was at a level 50% of the thigh length and half of the medio-lateral width of the VL. At this point, a longitudinal ultrasonic image of the VL was obtained (fig. 1) and recorded on a video tape which was synchronized with torque signals [Fukunaga et al., 1996]. At each joint angle, the torque was normalized to the maximal value and represented as %MVC.

From the ultrasound images at each torque level and at each joint angle the echoes from interspaces of fascicles and from the superficial and deep aponeuroses were visualized. The length of the fascicles was defined as the distance of ultrasonic echoes from the superficial to the deep aponeurosis measured by the use of a curvimeter (Comcurve-8, Koizumi, JPN). The fascicle angle was defined as the angle between the fascicles and the deep aponeurosis. The length and angle of the fascicles were measured five times in the same image, and three measurements excluding the largest and smallest values were averaged. The three measurements of fascicle lengths and fascicle angles did not differ significantly with the coefficient of variation (CV) ranging from 0 to 3.4% for the fascicle length and 0 to 7.5% for the fascicle angle.

Statistics

A two-way analysis of variance (ANOVA) was used to determine the effect of torque levels on fascicle lengths and fascicle angles at each knee joint angle. A tukey post hoc test was performed to test for significant differences among the means. Significance was accepted at $p<0.05$.

Results

The fascicle length related to knee joint angles and %MVC is shown in figure 2. When the torque increased from relaxed to MVC, fascicle length decreased curvilinearly with torque at each knee joint angle: more shortening was observed with the knee extended than flexed. At 100° and 10° of knee joint angles, fascicles shortened by 20 and 35%, respectively.

At each knee joint angle the fascicle angle increased with torque development, as shown in figure 3. Greater fascicle angles were observed, and fascicle angles increased more when the knee was extended (23% at 10° vs. 15% at 100°). The changes in fascicle length and angles due to torque development were greater at lower force levels (<30% MVC).

The active force-length characteristics of VL fascicles were estimated by the following method. It was assumed

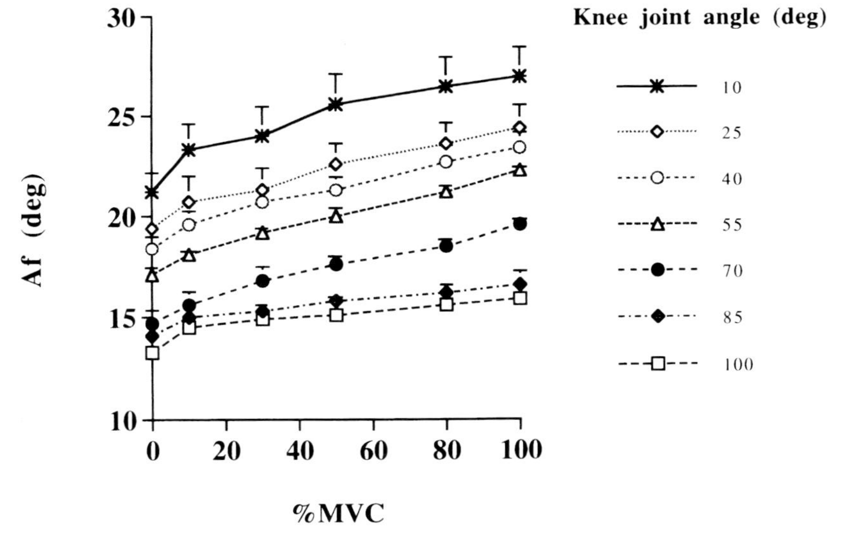

Fig. 3. The relationship between fascicle angles (Af) and knee extension torque (%MVC) at different knee joint angles. Symbols represent mean and SEM.

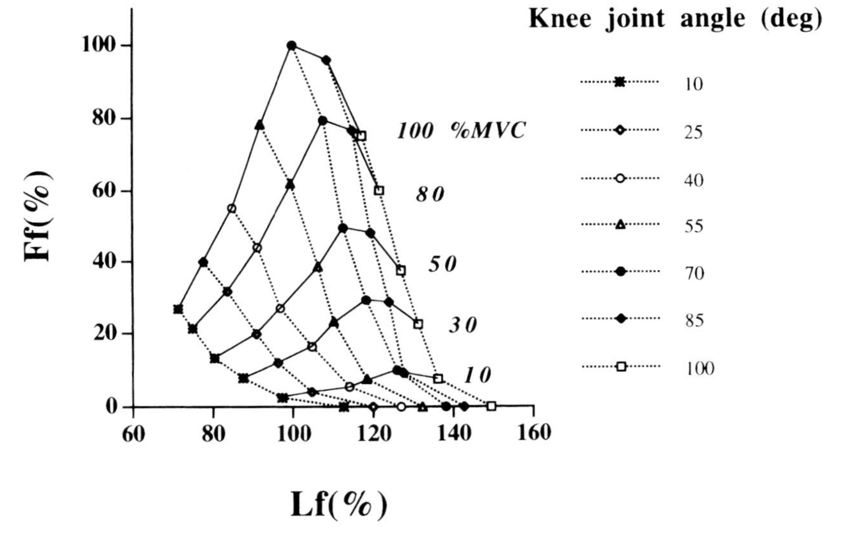

Fig. 4. The relationship between fascicle force (Ff, relative to the maximum value) and fascicle length (Lf, relative to the optimal length) at different knee angles, representing as mean values. The optimal length of fascicle was defined as the length at the maximum fascicle force. See text for calculation of fascicle force.

that the relative contribution of VL force to toal knee extension force was equal to the relative physiological cross-sectional area (PCSA) of this muscle at any knee joint angle and at any torque level. The VL force in the direction of the fascicle (Ff) was calculated as:

$$Ff = Ft \cdot K/\cos Af$$
$$Ft = TQ/MA$$

where Ft is the force at the tendon of quadriceps femoris, TQ is joint torque, MA is the moment arm length [Marshall et al., 1990], K is the relative PCSA of VL to the total PCSA of the quadriceps femoris (=34%: Akima et al. [1995]) and Af is the fascicle angle (°). The fascicle force reached a maximum (2,052 ± 125 N, mean ± SEM) at 78 mm (±2.6, SEM) of fascicle length which corresponded to a knee position of 70° where the fascicles had shortened by 27% compared with the relax condition. When the force exerted by the VL (normalized for the maximal force) was plotted against the relative fascicle length (fig. 4), the curve showed different patterns depending on the knee joint angles. A line joining the points at MVC force indicates the force-length curve during full activation of the muscle. The curve peaked when the knee was at 70°, and the Ff decreased with extension and flexion of the knee.

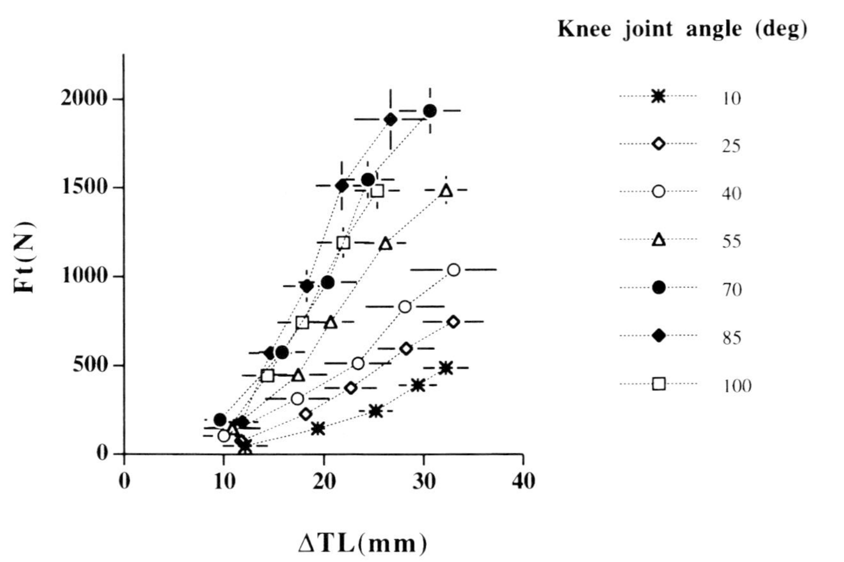

Fig. 5. The relationship between tendon elongation (ΔTL) and tendon force (Ft). See text for calculation of ΔTL and Ft. Symbols represent mean and SEM.

Discussion

The present technique was adequately reproducible for measurements of fascicle lengths and fascicle angles which is in good agreement with previous studies [Kawakami et al., 1993; Fukashiro et al., 1995; Herbert and Gandevia, 1995; Fukunaga et al., 1996; Narici et al., 1996; Fukunaga et al., 1997]. Therefore, the present ultrasonic methods can be used reliably to evaluate in vivo muscle architecture during contraction in human muscles.

The present study showed that fascicle length decreased with development of isometric force, and that this shortening was more with the knee extended than flexed. In the relaxed condition at a shorter muscle length, the muscle contractile component and the elastic component are slack [Huijing et al., 1989; Trestik and Lieber, 1993; Herbert and Gandevia, 1995]. It appears that during contraction, the slackness within the muscle was removed by fascicle shortening, and then the slackness of the tendon was taken up by further shortening of the fascicles which led to development of torque [Huijing et al., 1989]. A lengthening of the tendinous tissue (aponeurosis) would have been greater with the extended knee, since when the knee was flexed over 85°, fascicle shortening was almost linearly related with the increase in torque above 10% MVC, while when the knee was extended the relationship was curvilinear (fig. 2). Greater shortening of fascicles during weak contractions with the knee extended would reflect more taking-up of the slackness in the tendinous tissue.

The Ff-Lf curve during full activation of the muscle (MVC) indicated that the optimal length of a VL fascicle (78 mm) is observed at a knee position of 70° (fig. 4). In other words, during knee extension from 70°, the VL uses the ascending limb of the force-length relationship and when the knee is flexed over 70°, the VL operates on the descending limb of the force-length curve. Because the degree of taking-up of tendon slackness changes with muscle force, the optimal fascicle length differed at different activation levels (%MVC). As %MVC decreased, the force-length curves shifted to longer fascicle lengths. These results indicate that the relationship between Ff and fascicle length would change in a complex manner during daily physical activities, since these measurements are performed at various joint positions and involve various levels of muscle activation. The relationship between joint angles and fascicle lengths determined in a resting condition would be somewhat shifted from that of the actively contracting muscle [Herzog and ter Keurs, 1988].

The elongation of tendinous tissue (ΔTL) was estimated from the changes in fascicle length and fascicle angles during ramp isometric contractions by:

$$\Delta TL = Lf \cos Af - Lf' \cos Af',$$

where Lf, Lf′ and Af, Af′ are fascicle length and fascicle angles in relax and contracting conditions, respectively. When the tendon elongation (ΔTL) was related to the tendon force (Ft) (fig. 5), the relationship was curvilinear with less elongation at greater forces, and the force required to elongate a certain tendon length was greater when the knee was flexed. This relationship reflects the length-tension relationship of the tendinous tissue [Zuurbier et al., 1994] and would influence the active force-length characteristics of a muscle [Lieber et al., 1992]. Based on the length-tension relationship of tendon, the elongation should be larger when the overall length of the muscle-tendon unit is small (at an extended knee position). However, the maximal elongation was similar among the knee positions tested. This apparent discrepancy could be due to a smaller tendon force when the knee is extended.

In conclusion, muscle architectural parameters such as pennation angle and fascicle length change substantially during contraction, even when the contraction is 'isometric' and the length of the muscle-tendon unit is not changed. The architectural parameters should, therefore, be obtained for active muscles for more practical muscle models.

References

Akima, H., S. Kuno, T. Fukunaga, S. Katsuta (1995) Architectural properties and specific tension of human knee extensor and flexor muscles based on magnetic resonance imaging. Jpn J Phys Fitness Med *44:* 267–278.

An, K.N., F.C. Hui, B.F. Morrey, R.L. Linscheid, E.Y. Chao (1981) Muscle across the elbow joint: A biomechanical analysis. J Biomech *14:* 659–669.

Cutts, A. (1988) The range of sarcomere lengths in the muscles of the human lower limb. J Anat *160:* 79–88.

Friedrich, J.A., R.A. Brand (1990) Muscle fiber architecture in the human lower limb. J Biomech *23:* 91–95.

Fukashiro, S., M. Ito, Y. Ichinose, Y. Kawakami, T. Fukunaga (1995) Ultrasonography gives directly but noninvasively elastic characteristic of human tendon in vivo. Eur J Appl Physiol *71:* 555–557.

Fukunaga, T., M. Ito, Y. Ichinose, S. Kuno, Y. Kawakami, S. Fukashiro (1996) Tendinous movement of a human muscle during voluntary contractions determined by real-time ultrasonography. J Appl Physiol *81:* 1430–1433.

Fukunaga, T., Y. Ichinose, M. Ito, Y. Kawakami, S. Fukashiro (1997) Determination of fascicle length and pennation in a contracting human muscle in vivo. J Appl Physiol *82:* 354–358.

Griffiths, R.I. (1987) Ultrasound transit time gives direct measurement of muscle fibre length in vivo. J Neurosci Methods *21:* 159–165.

Herbert, R.D., S.C. Gandevia (1995) Changes in pennation with joint angle and muscle torque: in vivo measurements in human brachialis muscle. J Physiol *484:* 523–532.

Herzog, W., H.E.D.J. ter Keurs (1988) Force-length relation of in-vivo human rectus femoris muscles. Pflügers Arch *411:* 642–647.

Huijing, P.A., A.A.H. van Lookeren Campagne, J.F. Koper (1989) Muscle architecture and fibre characteristics of rat gastrocnemius and semimembranosus muscles during isometric contractions. Acta Anat *135:* 46–52.

Kawakami, Y., T. Abe, T. Fukunaga (1993) Muscle-fiber pennation angles are greater in hypertrophied than in normal muscles. J Appl Physiol *74:* 2740–2744.

Lieb, F.J., J. Perry (1968) Quadriceps function: An anatomical and mechanical study using amputed limbs. J Bone Joint Surg *50-A:* 1535–1548.

Lieber, R.L., C.G. Brown (1992a) Sarcomere length-joint angle relationships of seven frog hindlimb muscles. Acta Anat *145:* 289–285.

Lieber, R.L., C.G. Brown (1992b) Quantitative method for comparison of skeletal muscle architectural properties. J Biomech *25:* 557–560.

Lieber, R.L., C.G. Brown, C.L. Trestik (1992) Model of muscle-tendon interaction during frog semitendinosis fixed-end contractions. J Biomech *25:* 421–428.

Marshall, R.N., S.M. Mazur, N.A.S. Taylor (1990) Three-dimensional surfaces for human muscle kinetics. Eur J Appl Physiol *61:* 263–270.

Muhl, Z.F. (1982) Active length-tension relation and the effect of muscle pinnation on fiber lengthening. J Morphol *173:* 258–292.

Narici, M.V., G.C. Roi, L. Landoni, A.E. Minetti, P. Cerretelli (1989) Changes in force, cross-sectional area and neural activation during strength training and detraining of the human quadriceps. Eur J Appl Physiol *59:* 310–319.

Narici, M.V., T. Binzoni, E. Hiltbrand, J. Fasel, F. Terrier, P. Cerretelli (1996) In vivo human gastrocnemius architecture with changing joint angle at rest and during graded isometric contraction. J Physiol *496:* 287–297.

Rutherford, O.M., D.A. Jones (1992) Measurement of fibre pennation using ultrasound in the human quadriceps in vivo. Eur J Appl Physiol *65:* 433–437.

Trestik, C.L., R.L. Lieber (1993) Relationship between Achillis tendon mechanical properties and gastrocnemius muscle function. J Biomech Eng *115:* 225–230.

Wickiewicz, T., R.R. Roy, P.L. Powell, V.R. Edgerton (1983) Muscle architecture of the human lower limb. Clin Orthop Relat Res *179:* 275–283.

Woittiez, R.D., P.A. Huijing, R.H. Rozendal (1983) Influence muscle architecture on the length-force diagram of mammalian muscle. Pflügers Archiv *399:* 275–279.

Zuurbier, C.J., A.J. Everand, P. Van Der Wees, P.A. Huijing (1994) Length-force charateristics of the aponeurosis in the passive and active muscle condition and in the isolated condition. J Biomech *27:* 445–453.

Acta Anatomica

Acta Anat 1997;159:84–89

R.L. Lieber

Departments of Orthopaedics and Bioengineering, Biomedical Sciences Graduate Group, University of California and Veterans Administration Medical Centers, San Diego, Calif., USA

Muscle Fiber Length and Moment Arm Coordination during Dorsi- and Plantarflexion in the Mouse Hindlimb

Key Words
Fiber length
Sarcomere number
Force velocity
Length tension
Muscle architecture
Muscle mechanics

Abstract
The purpose of this study was to test the hypothesis that muscle fiber length and joint moment arm are combined in such a way that maximum muscle force is produced during locomotion. Plantarflexor (soleus, SOL and medial gastrocnemius, MG) and dorsiflexor (extensor digitorum longus, EDL and tibialis anterior, TA) muscle architecture in the mouse was measured along with their associated moment arms. Fiber length varied significantly between muscles ranging from 5.7±0.2 mm (MG) to 7.6±0.2 mm (TA). Plantarflexor moment arms were over twice as large as dorsiflexor moment arms (1.88±0.06 mm vs. 0.84±0.03 mm) suggesting a greater muscle length change with joint angle for plantarflexors compared to dorsiflexors. Using a simple muscle-joint model, the active sarcomere length range in these muscle groups was calculated and proved to be quite similar between functional groups. The active range for dorsiflexors was 2.2–2.4 μm, while the active range for plantarflexors was 2.2–2.5 μm, indicating that both muscle groups operate primarily near the plateau of their length tension-relation. Finally, when calculating force produced by muscles during locomotion, the combination of moment arm and fiber length measured in all muscle groups yielded muscle-joint systems that produced near maximal forces at the velocities modeled. These data indicate that fiber length and moment arm appear to be coordinated to yield the greatest possible force production during locomotion.

Introduction

Muscles generate the forces that rotate joints and cause animal locomotion. The torque and velocity requirements of a particular muscle-moment arm combination are defined by their role during normal movement, whether it be flying, swimming, hopping, manipulating or grasping. It would be naive to believe that the functional requirements of all muscle-joint systems would be the same under all of these conditions. However, it has been demonstrated that muscle shortening velocities during cat, frog and fish locomotion are in the range of 0.25–0.35 V_{max}, which is near the optimal velocity for power production [Walmsley et al., 1978; Altringham et al., 1993; Rome et al., 1993; Lutz and Rome, 1994]. This indicates that, across a variety of species and styles of locomotion, some muscles are used at velocities that result in maximal power production. Walking and trotting of digitigrade animals such as the mouse are characterized by slow ankle plantarflexion during the power stroke (E_3) followed by rapid dorsiflexion during the swing (F) phase [Goslow et al., 1973]. As such, it may not be surprising that dorsiflexors and plantarflexors show a number

KARGER
Fax +41 61 306 12 34
E-Mail karger@karger.ch
www.karger.com

Richard L. Lieber, PhD
Department of Orthopaedics, UC San Diego School of Medicine and VA Medical Center
3350 La Jolla Village Drive, La Jolla, CA 92093-9151 (USA)
Tel. +1 619 552 8585, ext. 7016, Fax +1 619 552 4381
E-Mail rlieber@ucsd.edu

of architectural differences in several species studied, such as the human [Wickiewicz et al., 1983], cat [Sacks and Roy, 1982], rabbit [Lieber and Blevins, 1989], guinea pig [Powell et al., 1984] and mouse [Burkholder et al., 1994]. These descriptions generally characterize the plantarflexors as muscles with relatively short fibers but large cross-sectional areas, an architectural specialization designed for force production, while the dorsiflexors are seen as muscles with relatively long fibers and small cross-sectional areas, an architectural specialization designed for velocity production.

Numerous studies have investigated the relationship between muscle architecture and function [Gans, 1982], but fewer have addressed the relationship between architecture and moment arm [Gans and Gaunt, 1991]. It is known that muscle fiber length and physiological cross-sectional area (PCSA) strongly influence maximum contraction velocity and isometric force, respectively. However, since the moment arm transforms the linear muscle motion into angular joint rotation, the moment arm influences both the muscle length range and shortening velocity required to produce a given joint motion. Thus, the interaction between muscle architectural properties and joint moment arms is critical for producing the correct joint kinematics during locomotion and may provide insights into the functional basis of muscle design. The purpose of this study was to investigate the coordination between muscle fiber length and moment arm in the mouse plantarflexor and dorsiflexor muscles as it relates to their normal function.

Materials and Methods

Architectural properties of the extensor digitorum longus (EDL), medial gastrocnemius (MG), soleus (SOL), and tibialis anterior (TA) muscles of female Swiss-Webster mice (n=6) were measured as described previously [Burkholder et al., 1994]. Briefly, hindlimbs were fixed in 10% buffered formalin with the knee and ankle held at 90°. Muscles were dissected, with special care to collect entire origins and insertions. Muscle mass, muscle length, and pennation angle were measured. Small fiber bundles were then teased from the muscle and mounted on microscope slides. The length of individual fibers was recorded under 60× magnification (Wild Model M3, Wild, Inc. Heerbrugg, Switzerland) to guarantee fiber removal from origin to insertion. Individual muscles were blotted dry, and muscle mass (M) determined using an analytical balance. Muscle length (L_m), defined as the distance from the origin of the most proximal muscle fibers to the insertion of the most distal muscle fibers, was determined using a calibrated digital filar eyepiece (Lasico Model 112983, Los Angeles, Calif., USA) and sarcomere lengths were measured by laser diffraction [Lieber et al., 1984]. Fiber length (L_f) was determined from individual fibers of each bundle using the filar eyepiece.

All muscle lengths and fiber lengths were then normalized to a sarcomere length of 2.5 μm to correct for variability induced by differences in the angle of joint fixation. Physiological cross-sectional

Table 1. Muscle and joint properties for dorsi- and plantarflexors

Muscle	Fiber length mm	PCSA mm^2	Moment arm mm
EDL	6.8±0.3	1.2±0.1	0.76±0.03
TA	7.6±0.2	5.4±0.2	0.89±0.04
MG	5.7±0.2	22.0±1.0	1.87±0.06
Soleus	6.4±0.4	1.2±0.06	1.90±0.06

Data obtained from n=4 (moment arm) or n=6 (muscle parameters) specimens. Values shown represent mean±SEM.

EDL=Extensor digitorum longus; TA=tibialis anterior; MG=medial gastrocnemius.

area (PCSA) was calculated as described by Sacks and Roy [1982] using the equation:

$$PCSA\ (mm^2) = \frac{M\ (g) \cdot \cos\theta}{\rho\ (g/mm^3) \cdot L_f\ (mm)}\ . \quad (1)$$

For moment arm determination, with the limbs freed of musculature, 7–0 polypropylene sutures were secured to the distal tendon stumps of dorsi- and plantarflexors. Then, the free end of the suture was led to the muscle origin, over a pulley and tied to a 3.5 gram mass. This mass was sufficient to maintain tension in the tendons, causing them to track along the normal muscle path during joint rotation, but was not so massive as to deform the joint and thus distort moment arm measurement. The ankle joint was then rotated through its range of motion, 50–150° (150° being full plantarflexion), while tendon motion was measured using the filar eyepiece. A line was fit to this excursion-joint angle record, and the slope taken as the average muscle moment arm about the ankle [An et al., 1983].

Architectural and moment arm data were then used to estimate isometric and dynamic joint torques mathematically. The geometric model used was similar to that developed by Zuurbier and Huijing [1992]. Muscle projection areas (A) was calculated as:

$$A = 2 \cdot L_m \cdot L_a \sin\beta, \quad (2)$$

where β is the included angle between the aponeurosis and the line of muscle action and L_a is aponeurosis length. Assuming the aponeurosis to be inextensible and the projection area to be constant, the aponeurosis angle can be calculated for any given muscle length from the projection area of the muscle in its reference state. The reference state for each muscle was calculated at sarcomere length 2.5 μm based on the measured architectural properties. The law of sines then yields the relationship between fiber length, muscle length and aponeurosis length as:

$$\frac{L_m}{\sin[180-(\alpha+\beta)]} = \frac{L_f}{\sin[\beta]} = \frac{L_a}{\sin[\alpha]}\ , \quad (3)$$

where α is the angle between the fibers and the muscle line of action. Equation 3 yields β and L_f from L_m, L_a, and α. The fiber velocity function was taken directly from Zuurbier and Huijing [1992]:

$$V_f = \frac{\cos(\alpha+\beta)}{\cos(\beta)} V_m, \quad (4)$$

where V_m is muscle velocity.

Table 2. Calculated sarcomere velocity and force reduction

Functional group	Muscle	Sarcomere velocity µm/s	Force reduction* %
Dorsiflexor	EDL	6.5	74
Dorsiflexor	TA	6.0	72
Plantarflexor	MG	1.3	29
Plantarflexor	soleus	1.2	37

* Force reduction resulting from shortening at the calculated sarcomere velocity.

Normalized muscle force-length and force-velocity properties were obtained from reasonable approximations in the literature. The force-length property was based on the relationship presented by Gordon et al. [1966] corrected for optimal force production at a sarcomere length of 2.5 µm; the force-velocity relationship was that presented by Katz [1939] with values for a and b obtained from the literature for mammalian muscle [Close, 1972]. The shortening force-velocity relation was described by:

$$F = \frac{V_{max} - V}{V_{max} + 4\,V} \quad (5)$$

V_{max} was taken to be 17.6 µm/s for fast fibers, and 9.0 µm/s for slow fibers. Fast and slow fibers were considered to have the same specific tension of 250 kPa.

Measured moment arms were then used to relate muscle lengths to joint angles. Joint kinematics during normal locomotion were taken from James et al. [1995]. Since the knee remained close to 90° during plantar- and dorsiflexion, it was held fixed throughout the model motions. Dynamic modeling was performed only during the time each muscle was activated, extrapolated using the EMG data of Nicolopoulos-Stournaras and Iles [1984]. Plantarflexion velocity was estimated as 220°/s, while dorsiflexion velocity was approximated as 1,300°/s [James et al., 1995], which represent the average angular velocities of shortening during a slow trot.

Results

Complete architectural results are presented elsewhere [Burkholder et al., 1994]; fiber lengths and PCSAs for the muscles discussed here are included in table 1 along with the appropriate values for moments arms. All linear regression fits to the excursion-joint angle were well approximated by the linear function as r^2 values always exceed 0.98.

Calculations based on measurement of sarcomere number and moment arm alone suggested that the dorsiflexor range of motion consisted primarily of the plateau region and ascending limb of the length-tension relation, where force is relatively constant, varying by 5–10% (fig. 1). The plantarflexor range of motion included a much larger portion of the ascending limb, indicating a decrease in force

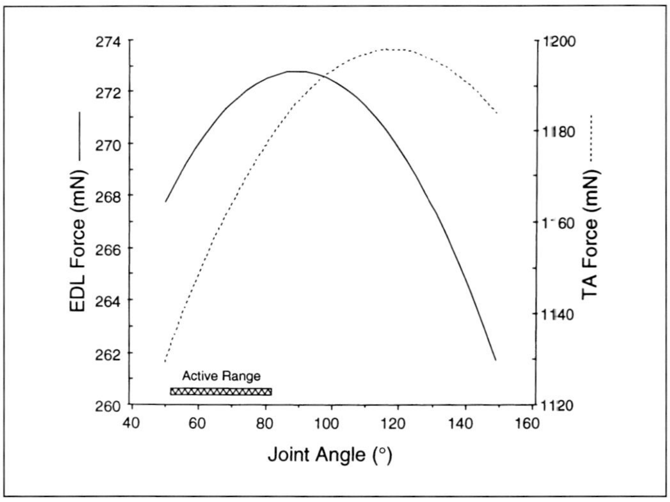

1

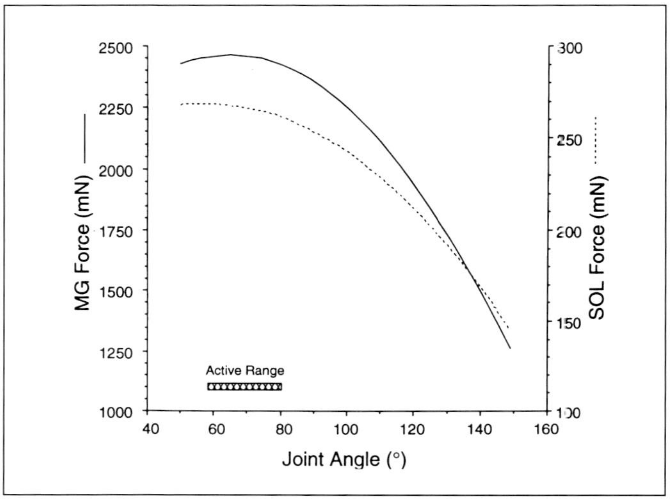

2

Fig. 1. Maximum isometric force-joint angle profiles calculated for both the EDL and TA based on measured fiber lengths and moment arms. Both muscles operate within 5–10% of their peak forces. The bar represents the joint range of observed during normal locomotion.

Fig. 2. Maximum isometric force-joint angle relations of the MG and soleus. Both muscles operate on the ascending limb of the force-length curve and demonstrate a much larger force variation throughout the joint range of motion compared to the TA and EDL. The bar represents the joint range of observed during normal locomotion.

potential with shortening which could vary by over 200% (fig. 2). Sarcomere lengths were calculated across the full anatomical range of motion, and were closely conserved within each functional group. MG and soleus had sarcomere lengths ranging from approximately 1.4–2.7 µm, while EDL and TA had shorter sarcomere length ranges, from 2.3 to 2.9 µm.

In contrast to the anatomically possible sarcomere length ranges, the active range of sarcomere lengths in these muscle groups was quite similar between functional groups (shaded bars, fig. 1, 2). The active range for dorsiflexors was approximately 2.2–2.4 μm (EDL: 2.30–2.48; TA: 2.15–2.33), while the active range for plantarflexors was 2.2–2.5 μm (MG: 2.25–2.56; SOL: 2.2–2.47). These data indicate that both muscle groups operate primarily near the plateau of their length tension relation.

Sarcomere velocity was also calculated to be similar within each of the two functional groups, though quite different between groups (table 2). The higher velocity of the dorsiflexor muscles reflected the higher angular velocity of dorsiflexion compared to plantarflexion, although the ratio of dorsiflexor sarcomere velocity to plantarflexor sarcomere velocity (5.5:1) was significantly less than their angular velocity ratio (10:1). This was a simple geometrical consequence of the greater fiber length:moment arm ratio of the dorsiflexors compared to the plantarflexors. The mouse ankle thus represents a muscle-joint system where differences in muscle architecture are partially offset by differences in joint kinematics.

Discussion

The most significant result of this study is that, within a functional group, moment arms and fiber lengths are matched to yield nearly identical sarcomere velocities. For the dorsiflexors, despite differences between TA and EDL muscle, fiber and moment arm lengths (table 1), active sarcomere shortening velocity in both muscles was roughly 0.35 V_{max} (table 2), very close to that predicted for optimal power generation. The conservation of sarcomere length range and shortening velocity in spite of architectural differences suggests a functional necessity of such a design.

The plantarflexors, in conjunction with the digital flexors are the primary muscles of digitigrade locomotion. They must be capable of the constant, low level activity associated with foraging, as well as the short bursts required for fleeing from predators. The former requires efficient production of low torques while the latter requires short bursts of great power. The dorsiflexors, on the other hand, act primarily in a nearly unloaded state to 'cock' the foot and toes during swing phase in preparation for the next step. As such, they must be capable of short bursts of rapid contractions against very low loads.

The fiber lengths and moment arms of these two muscle groups are combined in such a way so as to be suited for each of these particular tasks. The relatively short plantarflexor fiber lengths, and high pennation angles, particularly in the gastrocnemius, reflect a tendency toward packing sarcomeres in parallel to generate high force. The larger moment arm of the plantarflexors reflects a sacrifice of velocity for force production. To produce the same angular velocity, the plantarflexors would be required to shorten more than twice as fast as their antagonists. Although plantarflexion velocities may be quite large during the flight response, the normal daily activity of a caged laboratory mouse is better represented by the slower motions modeled here.

The dorsiflexors, with their longer fiber lengths and smaller moment arms reflect the alternate optimization –

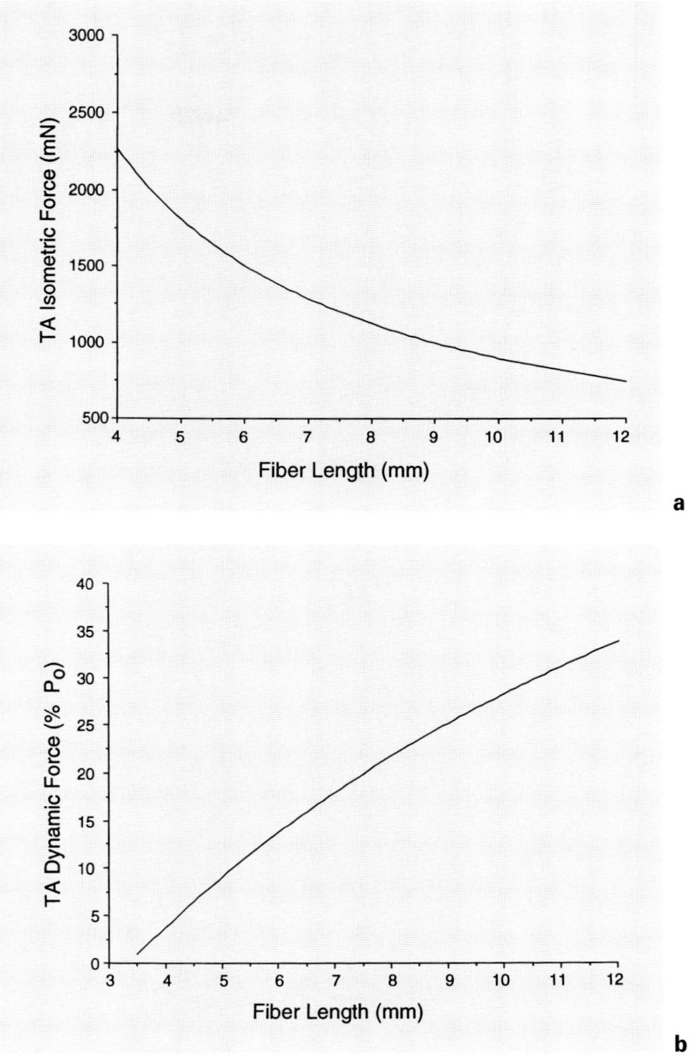

Fig. 3. **a** Theoretical relationship between muscle fiber length and isometric force for a constant muscle mass. **b** Theoretical relationship between muscle fiber length and dynamic muscle force at a constant shortening velocity.

4
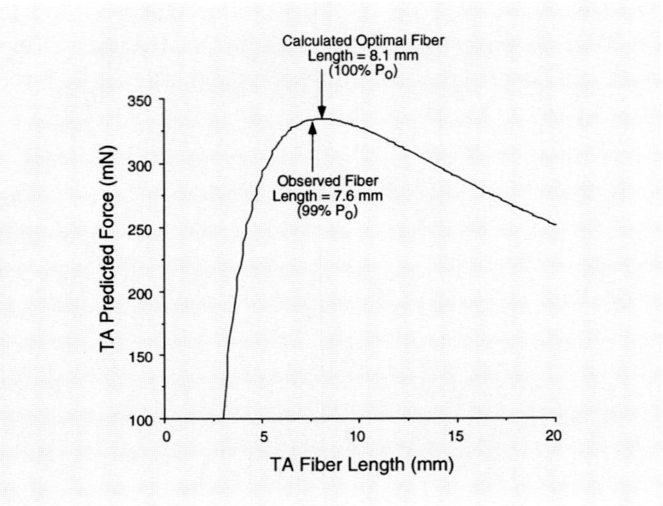

5
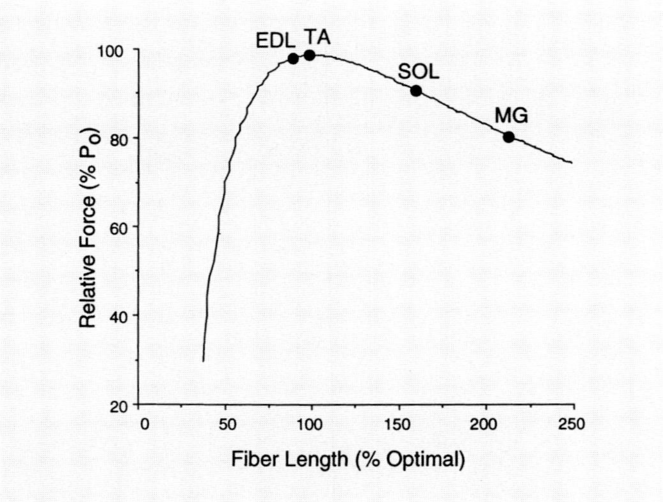

Fig. 4. Tension calculated to be generated by a TA muscle as fiber length is altered. Optimal fiber length results from a tradeoff between decreasing sarcomere shortening velocity (which increases force) and decreasing muscle isometric force capacity as fiber length increases.

Fig. 5. Tension calculated to be generated by the four muscles studied here as a function of normalized fiber length. Note that three of the four muscle generate near optimal force at the actual measured fiber length. The MG is an exception to this rule and may suggest that it is designed for normal use at a higher velocity than that used in the current simulation.

the sacrifice of force for speed. Both muscles are composed of long fibers, fairly parallel to the muscle axis, acting through moment arms that magnify angular rotation at the cost of mechanical advantage. The moment arm also permits these muscles to produce a great deal of joint rotation while maintaining near optimal sarcomere lengths. By virtue of the reduced linear velocity required for a given angular velocity, compared to the plantarflexors, the dorsiflexors benefit both in terms of their force-velocity and length-tension properties.

In theory, it is possible to arrange a given quantity of sarcomeres (i.e. a fixed muscle mass) to produce any combination of optimal muscle length and maximum shortening velocity simply by adjusting fiber length. This can be illustrated by changing the fiber length of the model TA from its observed fiber length of 7.6 mm to longer and shorter lengths while maintaining constant muscle length and muscle mass. The constraint of maintaining muscle mass, of course, requires a decrease in PCSA, and thus decreased isometric muscle force, with increasing fiber length (fig. 3a). In contrast, as fiber length increases, sarcomere velocity decreases for a given angular velocity, resulting in increased muscle force (fig. 3b). These two effects oppose one another resulting in a theoretically 'optimal' fiber length, representing the optimal simultaneous satisfaction of both criteria. By calculating the dynamic tension (i.e. tension during muscle shortening) as a function of calculated fiber length variation, the interaction between isometric force production (resulting from PCSA) and dynamic force production (sarcomere number and velocity) are seen (fig. 4). This presence of an optimal fiber length is the result of the fact that, at longer fiber lengths, PCSA continues to decrease hyperbolically, while sarcomere velocity decreases linearly. This diminishing sarcomere velocity increases force asymptotically towards P_O as fiber length increases but the concomitant reduced PCSA results in decreases in P_O with increasing fiber length that outweighs the effect of reduced shortening velocity. At shorter fiber lengths, the increased sarcomere shortening velocity overpowers the increased isometric force potential, again resulting in lower dynamic force production. For the mouse TA, dynamic tension peaked at a fiber length (8.1 mm) that was close to that observed by direct measurement (7.6 mm). The tension produced by the observed 7.6 mm fiber was still 99% of maximum.

This relationship was calculated for each muscle group studied and all muscles are plotted in figure 5 on a normalized fiber length scale for comparison. (The normalization value was the actual measured fiber length for each muscle.) For the EDL as with the TA, the observed fiber length was slightly shorter than that required for peak force production during our modeled motion. The MG, on the other hand, displayed fibers that were about two times longer length than would be optimal for this motion. The longer fiber lengths of the MG may indicate optimization for a faster motion. At faster gaits, the E_3 phase becomes faster, while the F phase remains relatively constant. As gait speed increases, the fiber length required for peak dynamic force production in the plantarflexors increases. Thus, the MG, which is activated at faster gaits, has fiber lengths that would be considered too long for the slow locomotion modeled here, and thus may reflect its use at higher velocities. The SOL muscle yielded a result intermediate between the 'optimized' TA and EDL and the MG. Whether this reflects the uniarticular nature of the SOL or its increased slow twitch fiber type percentage [Burkholder et al., 1994] remains to be determined.

In summary, this simulation suggests that muscle fiber length and moment arms of the mouse EDL, TA, MG, and soleus are well designed to produce the appropriate function during normal gait. Sarcomere lengths during shortening are appropriate for near peak isometric force production. Fiber length and moment arm appear to be coordinated to yield the greatest possible power production subject to the velocity requirements of the observed motion.

Acknowledgments

This work was supported by NIH grant AR31592. Thanks to Tom Burkholder for helpful comments and for performing many of the model calculations contained in this report.

References

Altringham, J.D., C.S. Wardle, C.I. Smith (1993) Myotomal muscle function at different locations in the body of a swimming fish. J Exp Biol *182:* 191–206.

An, K.N., Y. Ueba, E.Y. Chao, W.P. Cooney, R.L. Linscheid (1983) Tendon excursion and moment arm of index finger muscles. J Biomech *16:* 419–425.

Burkholder, T.J., B. Fingado, S. Baron, R.L. Lieber (1994) Relationship between muscle fiber types and sizes and muscle architectural properties in the mouse hindlimb. J Morphol *220:* 1–14.

Close, R.I. (1972) Dynamic properties of mammalian skeletal muscles. Physiol Rev *52:* 129–197.

Gans, C. (1982) Fiber Architecture and muscle function. Exercise and Sport Science Reviews. Lexington, Franklin University Press, vol 10, pp 160–207.

Gans, C., A.S. Gaunt (1991) Muscle architecture in relation to function. J Biomech *24S1:* 53–65.

Gordon, A.M., A.F. Huxley, F.J. Julian (1966) The variation in isometric tension with sarcomere length in vertebrate muscle fibres. J Physiol (Lond) *184:* 170–192.

Goslow, G.J., R. Reinking, D. Stuart (1973) The cat step cycle: Hind limb joint angles and muscle lengths during unrestrained locomotion. J Morphol *141:* 1–42.

James, R.S., J.D. Altringham, D.F. Goldspink (1995) The mechanical properties of fast and slow skeletal muscles of the mouse in relation to their locomotory function. J Exp Biol *196:* 491–502.

Katz, B. (1939) The relation between force and speed in muscular contraction. J Physiol (Lond) *96:* 45–64.

Lieber, R.L., F.T. Blevins (1989) Skeletal muscle architecture of the rabbit hindlimb: Functional implications of muscle design. J Morphol *199:* 93–101.

Lieber, R.L., Y. Yeh, R.J. Baskin (1984) Sarcomere length determination using laser diffraction. Effect of beam and fiber diameter. Biophys J *45:* 1007–1016.

Lutz, G.J., L.C. Rome (1994) Built for jumping: The design of the frog muscular system. Science *263:* 370–372.

Powell, P.L., R.R. Roy, P. Kanim, M. Bello, V.R. Edgerton (1984) Predictability of skeletal muscle tension from architectural determinations in guinea pig hindlimbs. J Appl Physiol *57:* 1715–1721.

Rome, L.C., D. Swank, D. Corda (1993) How fish power swimming. Science *261:* 340–343.

Sacks, R.D., R.R. Roy (1982) Architecture of the hindlimb muscles of cats: Functional significance. J Morphol *173:* 185–195.

Walmsley, B., J.A. Hodgson, R.E. Burke (1978) Forces produced by medial gastrocnemius and soleus muscles during locomotion in freely moving cats. J Neurophysiol *41:* 1203–1216.

Wickiewicz, T.L., R.R. Roy, P.L. Powell, V.R. Edgerton (1983) Muscle architecture of the human lower limb. Clin Orthop *179:* 275–283.

Zuurbier, C.J., P.A. Huijing (1992) Influence of muscle geometry on shortening speed of fibre, aponeurosis and muscle. J Biomech *25:* 1017–1026.

Acta Anatomica

Acta Anat 1997;159:90–98

Y. Ohira[a]
W. Yasui[a]
R.R. Roy[b]
V.R. Edgerton[b, c]

[a] Department of Physiology and Biomechanics, National Institute of Fitness and Sports, Kanoya City, Japan;
[b] Brain Research Institute, and
[c] Department of Physiological Science, University of California, Los Angeles, Los Angeles, Calif., USA

Effects of Muscle Length on the Response to Unloading

Key Words
Unloading
Muscle length changes
Immobilization
Fiber size
Succinate dehydrogenase activity
α-Glycerophosphate dehydrogenase activity
Myofibrillar ATPase activity

Abstract

Muscle mass, distribution of fiber types, fiber cross-sectional area (CSA) and selected enzyme activities were determined in rats hindlimb-suspended free of immobilization (Susp-Free), suspended with the ankle dorsiflexed (Susp-DF, soleus stretched) or plantarflexed (Susp-PF, soleus shortened) for 10 days and compared to cage-control (Con) rats. Reduction of muscle weight associated with suspension was prevented in Susp-DF rats. The mean CSAs of slow fibers were Con = Susp-DF > Susp-PF > Susp-Free and of fast and intermediate fibers tended to be Susp-DF > Con > Susp-PF = Susp-Free. Mean activities of succinate dehydrogenase (SDH), α-glycerophosphate dehydrogenase (GPD) and myofibrillar adenosine triphosphatase (mATPase) in slow and fast fibers were similar in Con and Susp-Free rats. Mean SDH activity in slow fibers was higher in Susp-DF and Susp-PF than in Con and Susp-Free. No significant differences in SDH activities of fast fibers were observed among groups. GPD activity was higher in slow fibers of Susp-DF and Susp-PF compared to Con. The mATPase activity was higher in slow fibers of Susp-DF compared to Con and Susp-Free rats and lower in fast fibers of Susp-DF compared to Con rats. Thus, when compared to control, the patterns of adaptation were more similar in the Susp-DF and Susp-PF than in the Susp-Free. Although these results are consistent with previous studies demonstrating that the load placed on a muscle can affect protein metabolism, the direction and magnitude of the adaptive responses observed in the present study were closely associated with the chronically imposed changes in muscle length, i.e. fixed at either a shortened or a lengthened position.

Introduction

A reduction in cross-sectional area (CSA) is induced in both slow and fast fibers of the rat soleus following gravitational unloading, although the magnitude of atrophy often is greater in the slow than in the fast fibers [Thomason and Booth, 1990; Roy et al., 1991; Ohira et al., 1992a; Edgerton and Roy, 1994, 1996; Roy et al., 1996a]. Unloading of antigravity muscles such as the soleus also appears to increase the percentage of fibers that display 'fast' characteristics.

Abbreviations used in this paper:
Con = Cage control; CSA = cross-sectional area; EMG = electromyography; GPD = α-glycerophosphate dehydrogenase; mATPase = myofibrillar adenosine triphosphatase; mATPase-Alk = myofibrillar adenosine triphosphatase staining at an alkaline pre-incubation; OD = optical density; SDH = succinate dehydrogenase; Susp-DF = hindlimb suspended with the ankle immobilized in a dorsiflexed position (soleus lengthened); Susp-Free = hindlimb suspended with the ankle not immobilized; Susp-PF = hindlimb suspended with the ankle immobilized in a plantarflexed position (soleus shortened).

KARGER

Fax +41 61 306 12 34
E-Mail karger@karger.ch
www.karger.com

0001–5180/97/1593–0090$15.00/0

This article is also accessible online at:
http://BioMedNet.com/karger

Yoshinobu Ohira, PhD
Department of Physiology and Biomechanics, National Institute of Fitness and Sports
Shiromizu 1, Kanoya City, Kagoshima Prefecture 891-23 (Japan)
Tel./Fax +81 994 46 4931
E-Mail ohira@nifs-k.ac.jp

Activities of mitochondrial enzymes measured in single fibers have been reported to be either unchanged or increased following periods of suspension or spaceflight ranging from 4 days to 4 weeks. Although there has not been a clear identification of the key factors which induce these adaptive changes in the mechanical and/or biochemical properties of slow extensor muscles, it is clear that the loading characteristics of the muscle is an important regulatory stimulus.

Unloading of the soleus results in the loss of muscle protein presumably due to a combination of slower protein synthesis and faster protein degradation rates [Thomason and Booth, 1990; Booth and Kirby, 1992]. For example, after about 1 week of hindlimb suspension the myofibrillar protein synthesis rates were decreased by ~60% and the degradation rates were increased by 7% in the soleus muscles of relatively nongrowing female rats [Thomason et al., 1989]. One factor which appears to have a profound effect on the protein metabolic rates during chronic unloading is the length at which the muscle is maintained. For example, Goldspink et al. [1986] and Jaspers et al. [1988, 1989] have demonstrated that suspension-associated atrophy of the soleus muscle can be prevented by maintaining the muscle in a stretched position by immobilizing the ankle in a dorsiflexed position. Apparently chronic stretch of an unloaded muscle results in an elevated protein synthesis above control levels and a reduction of the elevated protein degradation rates associated with unloading [Jaspers et al., 1988]. In addition, some of the characteristic adaptations in amino acid metabolism associated with unloading are diminished in the suspended-stretched soleus muscle [Jaspers et al., 1989]. Interestingly, if the soleus muscle was maintained in a shortened position during suspension for 6 days, the loss in mass, amino acid metabolism and the net protein balance of the soleus was similar to that in the contralateral (not immobilized) muscle [Jaspers et al., 1985].

The activation level of the soleus muscle is affected by both hindlimb suspension and immobilization. Alford et al. [1987] have shown that the total integrated electromyographic (EMG) activity per day decreases dramatically the first day of suspension, but begins to recover at 3 days and is near control levels after 7 days of suspension. During a period of limb immobilization, the daily activation levels of the soleus are dependent on the length at which it is immobilized: the highest activity level is associated with a lengthened muscle and the lowest with a shortened muscle [Fournier et al., 1983]. Thus, in hindlimb-suspended rats, the soleus immobilized in a lengthened position would be expected to show a higher activation level than when immobilized in a shortened position. Since the limb is fixed, it is presumed that isometric forces would be produced during activation and these active forces would be greater in the lengthened muscle. In addition, the passive forces would be expected to be larger in the lengthened compared to shortened muscles. On the other hand, a complicating factor in each of the experiments in which the muscle was chronically maintained at a given length is that the actual load on the muscle throughout the experiment has not been monitored. This load could change because fiber lengths, and thus the number of sarcomeres in series per fiber, can change within a few days of a chronic change in length [Goldspink et al., 1977; Spector et al., 1982].

The purpose of the present study, therefore, was to determine the effects of immobilizing the unloaded soleus at different lengths (ankle dorsiflexed-soleus lengthened; or ankle plantarflexed-soleus shortened) on the size and metabolic properties of its fibers. The data suggest that the level of mechanical loading associated with the active and passive forces placed on a muscle by varying its length is an important determinant of the atrophic response. Further, the data indicate that chronic stretching of an unloaded muscle during hindlimb suspension ameliorates many of the adaptations normally associated with chronic unloading.

Materials and Methods

Experimental Design and Hindlimb Suspension Procedure

All animal care and experimental procedure were performed according to the institutional guidelines. Eighteen adult male Wistar rats (Kyudo, Kumamoto, Japan) weighing ~250 g were separated randomly into: (1) ambulatory cage control (Con); (2) hindlimb suspended with the ankle joints not immobilized (Susp-Free); and (3) hindlimb suspended with the left ankle in a dorsiflexed (Susp-DF) and the right ankle in a plantarflexed (Susp-PF) position. Food and water were supplied ad libitum. All animals were housed individually in stainless steel cages (29×50×20 cm). The temperature and humidity in the animal room with 12:12 h of light:dark cycle were maintained at ~23 °C and ~55%, respectively.

To suspend the hindlimbs, a modification of the method described by Morey [1979] was used. Briefly, a harness was placed loosely around the lower trunk to aid in the support of the casted legs (see below). The suspended rats were able to eat food and drink water freely by using their forelimbs. Rats selected for the immobilization group were lightly anesthetized during the immobilization procedures. Using plaster casts, the left ankle joint was fixed at ~50° of dorsiflexion, thereby chronically stretching the soleus. The right joint was fixed at ~160° of plantarflexion, thereby chronically shortening the soleus. Hindlimb suspension was maintained for 10 continuous days. To avoid the effects of loading the limbs due to the plaster casts, the legs were supported (knee at ~90°) via a suture from the cast to the harness.

After 10 days of suspension with or without joint immobilization, the rats were anesthetized with sodium pentobarbital (5 mg/100 g body weight i.p.) while they were still suspended. The plaster casts were removed carefully, the rats were weighed and the ankle joint angles measured. Soleus muscles were removed from both limbs, cleaned of fat and connective tissue and weighed (wet weight). The midportion of the soleus was mounted on a cork using OCT compound (Miles, Inc., Ind., USA) and immediately frozen in isopentane cooled in liquid nitrogen. Only the left soleus was used for the histochemical analyses in the cage controls and suspended group without joint immobilization. Muscle samples were stored at –80 °C until analyzed.

Fiber Typing Procedures

Serial cross-sections (10 μm thickness) were cut in a cryostat at –20 °C and stained qualitatively for myofibrillar adenosine triphosphatase (mATPase) at an alkaline (pH 8.75) preincubation (mATPase-Alk) as described by Nwoye et al. [1982]. Tissue sections were digitized as gray level pictures on a computer-enhanced image processing system and were stored on magnetic tapes as described in detail elsewhere [Martin et al., 1985]. Briefly, arrays of picture elements were quantified to 256 gray levels that were then converted automatically to an optical density (OD). A population of ~65 fibers from the midbelly of each muscle was analyzed for CSA. The staining reaction of mATPase-Alk was used to categorize each fiber as light (slow), dark (fast) or intermediate on the basis of OD readings. For each animal, the separation into three subpopulations was very distinct.

Quantitative Histochemical Procedures

The activities of succinate dehydrogenase (SDH), an oxidative marker enzyme, and α glycerophosphate dehydrogenase (GPD), a glycolytic marker enzyme, were determined histochemically as described previously [Martin et al., 1985; Jiang et al., 1992; Ohira et al., 1992a]. A modification [Jiang et al., 1992] of the histochemical methods of Weisberg et al. [1982] was used to measure mATPase activity. For each assay, three sections were incubated in a medium that contained substrate, and three sections were incubated in a medium that lacked substrate. Sections incubated without the substrate were used as tissue blanks to correct for nonspecific staining that occurred during the reaction. The activities were expressed as ΔOD/min and as integrated OD (ΔOD/min × fiber CSA). Each fiber then was matched with mATPase-Alk staining for fiber typing. Fascicles of fibers free of tissue artifact were chosen for all analyses.

Statistical Procedures

The reported data are expressed as the mean ± SEM. The data were analyzed using a one-way analysis of variance followed by a Student's t test. The 0.05 level was established for statistical significance.

Results

Muscle Weight

Both the mean absolute and relative (to body weight) wet weights of the soleus in Susp-Free and Susp-PF groups were smaller than in Con rats (fig. 1). In the Susp-DF group, the absolute weight was similar to Con and the relative weight was larger than Con. In addition, both the absolute and relative weights of the Susp-DF group were larger than in the Susp-Free and the Susp-PF groups.

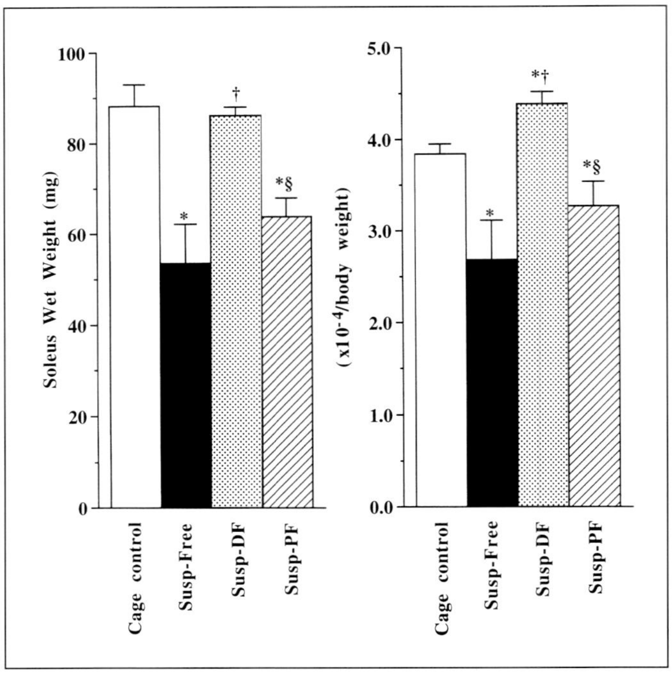

1

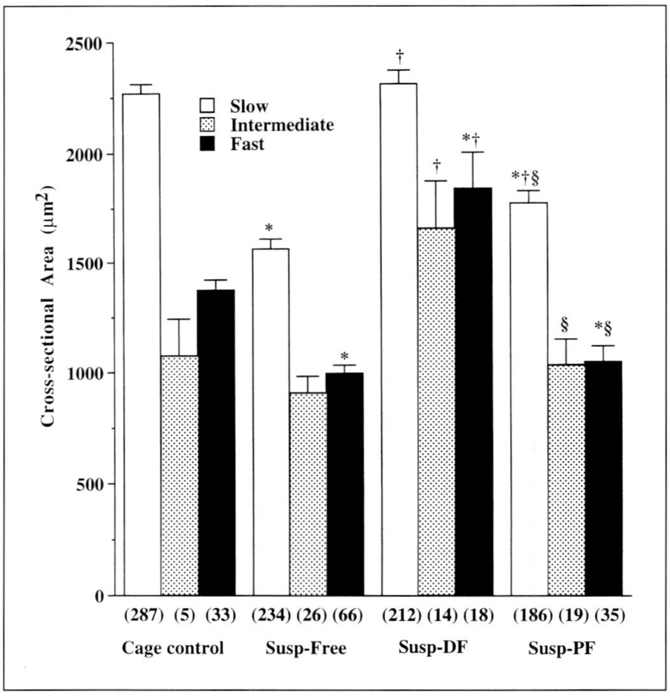

2

Fig. 1. Absolute and relative wet weights (means ± SEM) of the soleus. Susp-Free, Susp-DF and Susp-PF: hindlimb-suspended with the ankle joint in a free position, or immobilized in a dorsiflexed or plantarflexed position, respectively. *, †, § Significantly different from cage control, Susp-Free or Susp-DF at $p<0.05$.

Fig. 2. Cross-sectional areas (mean ± SEM) of soleus fibers. The numbers of fibers analyzed for each group are shown in parentheses. Abbreviations are the same as in figure 1. *, †, § Significantly different from cage control, Susp-Free or Susp-DF at $p<0.05$.

Fiber Size

The adaptations in fiber CSA reflected the changes in muscle wet weights (fig. 2). The mean CSA of slow fibers was Con = Susp-DF > Susp-PF > Susp-Free. The mean CSAs of the intermediate and fast fibers were Susp-DF > Con > Susp-Free = Susp-PF.

Fiber Type

Con rats contained ~90% slow fibers and this was maintained in the Susp-DF rats (fig. 3). The percentage of slow fibers was reduced to ~70% ($p<0.05$) and 75% ($p>0.05$) in the Susp-Free and Susp-PF rats, respectively. Both the Susp-Free and Susp-PF rats showed small increases in the percentages of fast and intermediate fibers compared to Con. However, only the percentage of intermediate fibers was significantly higher in the Susp-PF than the Con rats. In addition, the percentage of fast fibers in the Susp-DF was lower compared to the Susp-Free rats.

Succinate Dehydrogenase Activity

Compared to Con, the mean SDH activity of the slow fibers was unchanged in Susp-Free rats and elevated in both immobilized groups (fig. 4). Mean SDH activities of the fast and intermediate fibers in the suspended groups were similar to Con. However, compared to Susp-Free rats the mean SDH activities of the intermediate fibers were higher in the Susp-DF and Susp-PF rats. The fast fibers had similar mean SDH activities across all groups. When the changes in CSA were considered, the mean total SDH activity per fiber in both slow and fast fibers was lower in Susp-Free than in Con rats. The slow fibers in the Susp-DF rats had a higher total SDH activity than the slow fibers in Con rats. Each fiber type in both the Susp-DF and Susp-PF rats showed elevated activities compared to Susp-Free rats. The fast fibers in the Susp-PF rats had lower total SDH activities than the Con rats. In general, these results demonstrate a net increase of SDH enzyme in the soleus of the Susp-DF and Susp-PF rats relative to the Susp-Free rats.

α-Glycerophosphate Dehydrogenase

The mean GPD activity in all three types of fibers tended to be higher in the Susp-Free compared to the Con group ($p>0.05$) (fig. 5). In the Susp-DF and Susp-PF groups, the mean GPD activity of the slow fibers was higher than in Con rats. The total activity of GPD per fiber was significantly elevated in both the slow and fast fibers of the Susp-Free compared to the Con rats. In the Susp-DF and Susp-PF groups, the total GPD activity was increased only in the slow fibers relative to Con. The total GPD activity of the slow fibers of the Susp-DF group was elevated compared

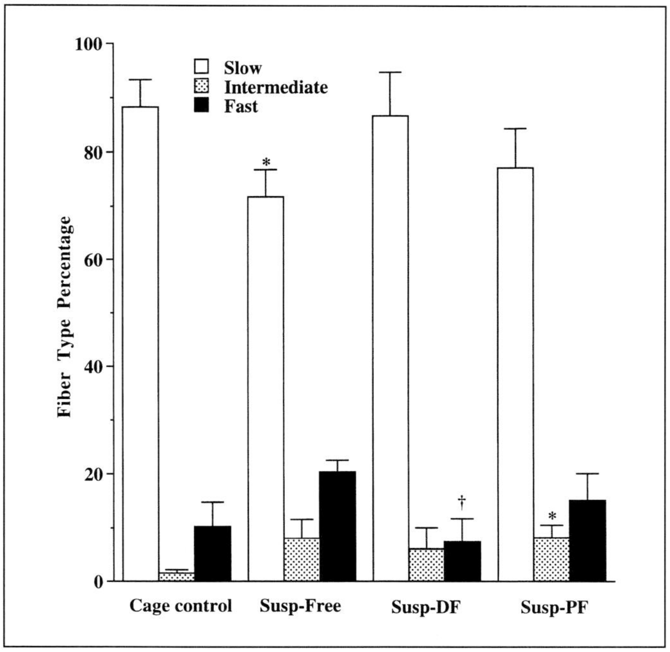

3

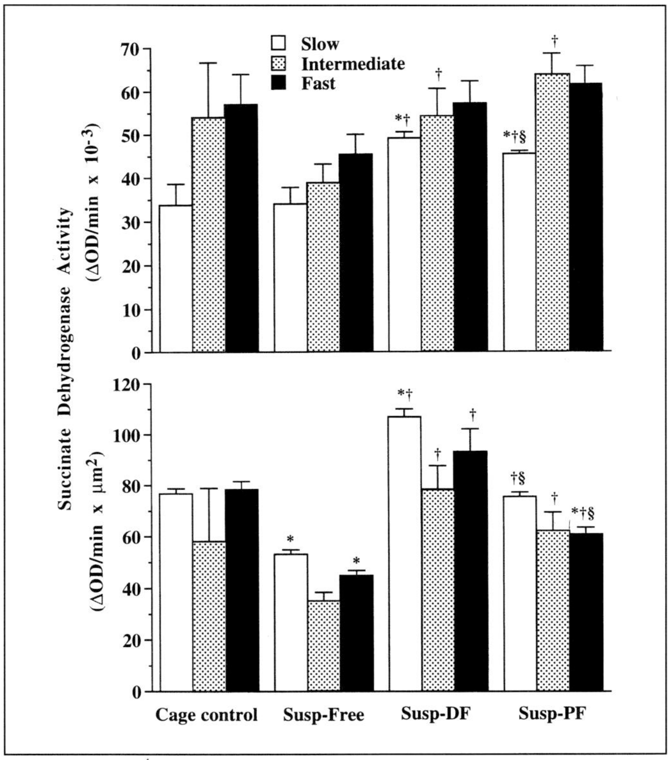

4

Fig. 3. Fiber-type composition (mean ± SEM) of the soleus. Abbreviations are the same as in figure 1 and numbers of fibers analyzed are the same as in figure 2. *, † Significantly different from cage control, Susp-Free at $p<0.05$.

Fig. 4. Succinate dehydrogenase activities (mean ± SEM) of soleus fibers. Abbreviations are the same as in figure 1 and numbers of fibers analyzed are the same as in figure 2. *, †, § Significantly different from cage control, Susp-Free or Susp-DF at $p<0.05$.

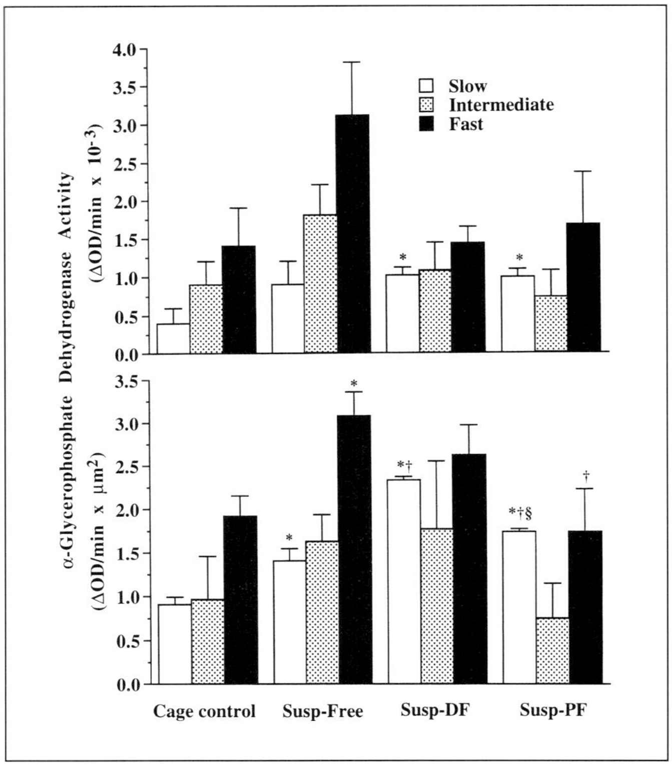

Fig. 5. α-Glycerophosphate dehydrogenase activities (mean ± SEM) of soleus fibers. Abbreviations are the same as in figure 1 and numbers of fibers analyzed are the same as in figure 2. *, †, § Significantly different from cage control, Susp-Free or Susp-DF at $p<0.05$.

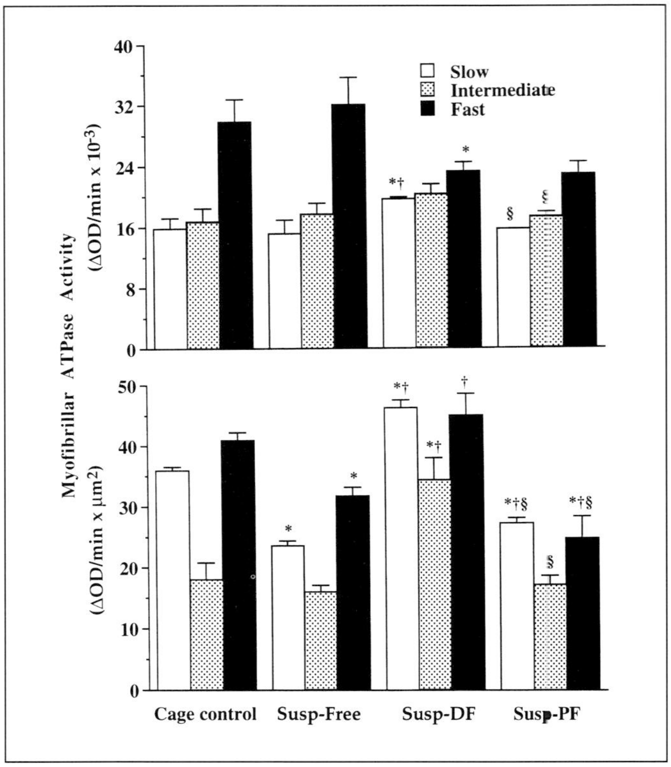

Fig. 6. Myofibrillar ATPase activities (mean ± SEM) of soleus fibers. Abbreviations are the same as in figure 1 and numbers of fibers analyzed are the same as in figure 2. *, †, § Significantly different from cage control, Susp-Free or Susp-DF at $p<0.05$.

to the Susp-Free and Susp-PF groups. These results show higher GPD enzyme units per fiber in the slow fibers of all suspended groups compared to Con in spite of the fact that the highest level of atrophy occurred in the slow fibers of all groups except for the Susp-DF group.

Myofibrillar Adenosine Triphosphatase

Suspension with the ankle in the 'free' position had no effect on mATPase activity (fig. 6). Suspension in the DF position (soleus stretched) resulted in an increase in the mATPase activity in the slow and a decrease in the fast fibers compared to Con. The mATPase activities of each fiber type of the Susp-PF group were similar to control while those of the slow and intermediate fibers were lower than in Susp-DF rats. Thus, the mATPase activity of the slow fibers appears to have been elevated by stretching the soleus muscle, i.e. mATPase activities in Susp-DF > Susp-PF and Susp-Free. The total activities of mATPase per fiber in both slow and fast fibers were lower in the Susp-Free and Susp-PF than in the Con group. In contrast, stretching the muscle (Susp-DF group) resulted in an elevation of the total mATPase activity in the slow and intermediate fibers and no change in the fast fibers relative to control levels. Further, the activities in each fiber type were higher in Susp-DF compared to Susp-Free and Susp-PF rats. These results show that mATPase enzyme units are markedly increased when the soleus is suspended in a stretched position.

Discussion

The range of movement of the rat ankle joint is ~30 to 180°, but the excursion at the ankle in cage-control rats is generally maintained between ~30 to ~90° [Ohira et al., 1992b, 1993]. Although muscle or fiber lengths were not measured in the present study, the length of the soleus, excluding the tendon, of male Wistar rats (n = 5, mean body weights of 318 g) suspended for 10 days was 28 and 20 mm when the ankle was immobilized at an angle of ~50° compared to ~160°, respectively [Ohira and Yasui, unpubl. observations]. These data are consistent with other studies involving chronic immobilization of the rat soleus muscle at either a shortened or lengthened state [Spector et al., 1982]. Interestingly, Roy et al. [1996b] reported no change in muscle length but a significant increase (~10%) in the mean fiber length in the soleus muscle of adult rats suspended for 28 days with the ankle in a 'free' position. Similar data have been reported for young rats suspended for either 10 or 30 days [Darr and Schultz, 1989]. Thus, it appears that when the hindlimb is suspended with the ankle in a 'free' position, the length changes associated with activation maintains soleus muscle length. The increase in mean fiber lengths may be associated with a rearrangement of the fibers and or connective tissue in 3-D as the muscle severely atrophies.

The mean wet weight of the soleus fixed at a shortened muscle length (Susp-PF) or not fixed (Susp-Free) during suspension was significantly less than that in the muscles fixed in a lengthened position (Susp-DF) or in the cage controls. In general, the relative changes in CSA of both slow and fast fibers were altered similarly in response to muscle length changes. The only exception was a significant increase in the mean size of the slow fibers but no change of the fast fibers in Susp-PF compared to Susp-Free rats. The CSA of the fast fibers in the Susp-DF was even greater than in the muscles of the Con rats. Together these results suggest that the fast fibers are more responsive at the higher levels of stretch and less sensitive to the lower levels of stretch than the slow fibers. In contrast, the slow fibers are less responsive at the most stretched position and more responsive at the shorter lengths compared to fast fibers. The Susp-Free position had a greater negative effect on the size of the slow fibers than even the shortened fixed position. The conclusion from these results is that the chronic passive length imposed on a fiber affects CSA, but regulatory factors associated with muscle length other than the apparent load may also influence the magnitude or even the direction of the adaptation. It is conceivable that the forces generated in the Susp-Free position could be less than in the Susp-PF position, particularly if there was a loss of serial sarcomeres in the shortest position. For example, at any given level of activation the muscle in a 'fixed' position can exert more maximal, i.e. isometric, force than it can when the muscle is allowed to shorten with minimal load. The present data are consistent with the findings of Leterme et al. [1994] who reported a maintenance of soleus muscle mass, maximum tetanic tension and fiber type distribution in 14-day hindlimb suspended soleus muscles that were chronically stretched. However, the mean size of each fiber type remained significantly smaller than control in the study by Leterme et al. [1994], whereas the fibers were at or above control levels in the present study. The reason for these different results is unknown.

On the other hand, these results cannot determine with certainty that there is a direct relationship between the fixed muscle length and the level of chronic loading since the active tension was not measured nor estimated from EMG recordings in the present study. We have reported that the intramuscular EMG in a slow and a fast rat ankle extensor, i.e. the soleus and medial gastrocnemius, disappears instantaneously following hindlimb suspension [Alford et al., 1987]. However, the activation appears to recover to pre-suspension control levels after 7–10 days of suspension [Alford et al., 1987; Ohira et al., 1992b]. It also appears that most of the muscle atrophy has already occurred during this 7- to 10-day period of reduced activation [Thomason and Booth, 1990; Edgerton and Roy, 1996]. The EMG activity in the soleus of the Susp-DF group is sustained during suspension, although the magnitude of EMG gradually declines [Ohira et al., 1992b]. The EMG activity in the shortened soleus was initially reduced and then gradually recovered to presuspension levels [Ohira et al., 1992b], resembling that seen in the soleus of rats suspended without any joint immobilization [Alford et al., 1987]. In any case, it would seem that minimal tension would be produced in the muscle fixed at a shortened length even when EMG is present [Ohira et al., 1992b; Ohira and Edgerton, 1994]. Further, Fournier et al. [1983] reported that the total quantity of EMG during a 24-hour period was not a major factor that could be associated with the changes in the functional properties of skeletal muscles immobilized at different lengths in normal rats. It is also known that passive muscle stretching and/or development of tension by stretching increases protein synthesis [Goldspink et al., 1986; Loughna et al., 1986; Jaspers et al., 1988]. These data combined with the present results suggest that either loading or tension development, independent of electrical activity, can be an important factor in the maintenance of muscle mass in vivo.

There are numerous reports of some slow fibers in the soleus shifting to a fast type following unloading [Thomason et al., 1987; Thomason and Booth, 1990; Roy et al., 1991; Ohira et al., 1992a; Edgerton and Roy, 1994, 1996; Roy et al., 1996a]. A reduction from ~90 to ~70% ($p<0.05$) or ~75% ($p>0.05$) slow fibers after only 10 days of suspension with the soleus in the 'free' or shortened position was observed in the present study. Since this change in fiber types in response to suspension was present when the soleus was in a shortened position with or without joint immobilization, it appears that one of the factors influenc-

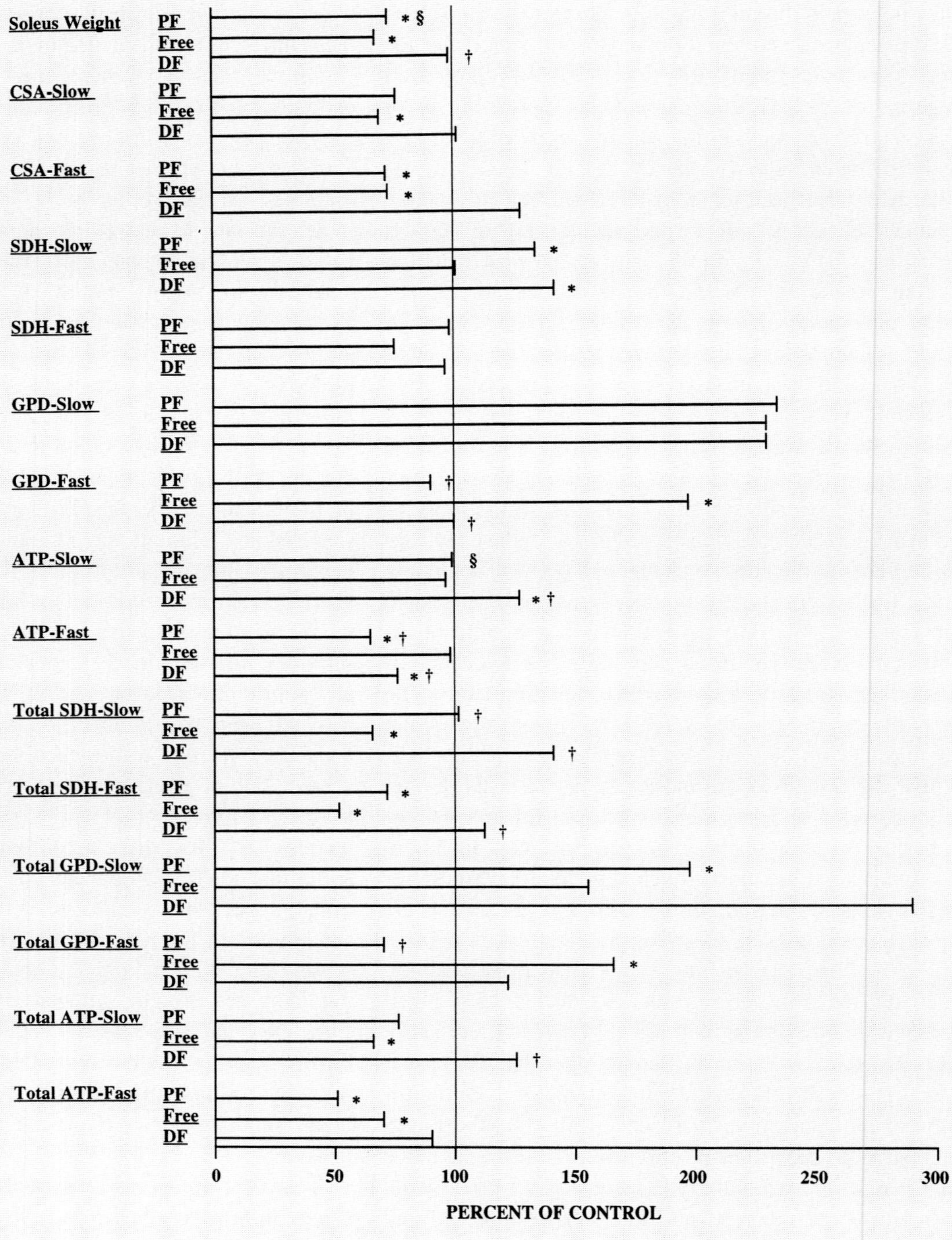

Fig. 7. Summary of the mean percent values relative to the cage-control values for each parameter in the soleus muscle of hindlimb-suspended rats with their ankle joints in a free position (Free), or immobilized in a dorsiflexed (DF) or plantarflexed (PF) position. The vertical line at 100% represents the control values. CSA = Cross-sectional area; SDH = succinate dehydrogenase; GPD = α-glycerophosphate dehydrogenase; ATP = myofibrillar adenosine triphosphatase; slow and fast = slow and fast fibers, respectively. *, †, § Significantly different from cage control, Free or DF at $p<0.05$.

ing some fibers to convert toward a fast myosin composition is a reduction in chronic force.

The present results, however, also demonstrate that the modulation of the size and metabolic properties of slow and fast fibers is a response to a change in the chronic length of the muscle, whether the change is to a lengthened or shortened state. For example, the values for most muscle properties measured were more similar when the soleus was in the shortened or lengthened position compared to when it was in the neutral ('free') position. These relative relationships are summarized in figure 7. For example, muscle mass and fiber size were reduced more in the Susp-Free than in either the Susp-DF or Susp-PF rats. A similar response was evident for the decrease in the percent slow fibers (fig. 3) and for most of the adaptations in enzyme activities (fig. 4–7). For example, the mean and total SDH activities of both slow and fast fibers were the lowest in the Susp-Free group. The mean and total GPD activities were elevated the least in the slow fibers and the most in the fast fibers compared to control. Thus, for fast fibers the expected elevation in GPD activity with suspension was blunted by changing the muscle length, while for the slow fibers the activity was enhanced. In no case, except for total mATPase activities for the fast fibers, were the relative effects among groups in the order of Susp-DF, Susp-Free and Susp-PF as would be expected if the modulating variable was muscle length only. A previous study [Bruce-Gregorios and Chou, 1984] reported the appearance of core fibers, i.e. fibers characterized by distinct zones of myofibrillar structure, in the soleus following 1–3 weeks of immobilization in a shortened position. The appearance of such fibers could have complicated the interpretation of the present quantitative histochemical results since these zones of myofibrillar disruption would have affected staining intensities. However, no core fibers were observed in the present study.

Perspective

The conclusion that the change in muscle length, not the length of the muscle per se, influenced the direction and magnitude of the adaptations suggests that the stimulus to adjust fiber lengths by adding or reducing the number of sarcomeres in series induces similar adaptive responses. These adaptive responses associated with a lengthening or shortening of the muscle fiber length, combined with the presumed ability to exert load when fixed at either position, appear to differ from the condition which does not induce a change in the length of the fibers, i.e. the number of sarcomeres presumably do not change nor can significant load be generated by the muscle. Further, the present results highlight the need of carefully monitoring force and EMG signals during experimental treatments designed to modify load, activation and use-related variables.

Acknowledgments

This study was supported by the Grant-in Aid for Scientific Research (B) from the Ministry of Education, Science, Sports and Culture of Japan and by funds from the Institute of Space and Astronautical Science and Japan Space Utilization Promotion Center. The authors thank Jung Kim and Chau Tri for their expert assistance in preparing the figures.

References

Alford, E.K., R.R. Roy, J.A. Hodgson, V.R. Edgerton (1987) Electromyography of rat soleus, medial gastrocnemius, and tibialis anterior during hind limb suspension. Exp Neurol *96:* 635–649.

Booth, F.W., C.R. Kirby (1992) Changes in skeletal muscle gene expression consequent to altered weight bearing. Am J Physiol *262:* R329–R332.

Bruce-Gregorios, J., S.M. Chou (1984) Core myofibers and related alterations induced in rats' soleus muscle by immobilization in shortened position. J Neurol Sci *63:* 267–275.

Darr, K.C., E. Schultz (1989) Hindlimb suspension suppresses muscle growth and satellite cell proliferation. J Appl Physiol *67:* 1827–1834.

Edgerton, V.R., R.R. Roy (1994) Neuromuscular adaptation to actual and simulated weightlessness; in Bonting SL (ed): Advances in Space Biology and Medicine. Greenwich, JAI Press, vol 4, pp 33–67.

Edgerton, V.R., R.R. Roy (1996) Neuromuscular adaptations to actual and simulated spaceflight; in Fregly MJ, CM Blatteis (eds): Handbook of Physiology, sect 4. Environmental Physiology. III. The Gravitational Environment, chapt 32. New York, Oxford University Press, pp 721–763.

Fournier, M., R.R. Roy, H. Perham, C.P. Simard, V.R. Edgerton (1983) Is limb immobilization a model of muscle disuse? Exp Neurol *80:* 147–156.

Goldspink, D.F. (1977) The influence of immobilization and stretch on protein turnover of rat skeletal muscle. J Physiol (Lond) *264:* 267–282.

Goldspink, D.F., A.J. Morton, P. Loughna, G. Goldspink (1986) The effect of hypokinesia and hypodynamia on protein turnover and the growth of four skeletal muscles of the rat. Pflügers Arch *407:* 333–340.

Jaspers, S.R., J.M. Fagan, M.E. Tischler (1985) Biochemical responses to chronic shortening in unloaded soleus muscle. J Appl Physiol *59:* 1159–1163.

Jaspers, S.R., J.M. Fagan, S. Satarug, P.H. Cook, M.E. Tischler (1988) Effects of immobilization on rat hind limb muscles under non-weight-bearing conditions. Muscle Nerve *11:* 458–466.

Jaspers, S.R., E.J. Henriksen, S. Satarug, M.E. Tischler (1989) Effects of stretching and disuse on amino acids in muscles of rat hind limbs. Metabolism *38:* 303–310.

Jiang, B., Y. Ohira, R.R. Roy, Q. Nguyen, E.I. Ilyina-Kakueva, V. Oganov, V.R. Edgerton (1992) Adaptation of fibers in fast-twitch muscles of rats to spaceflight and hindlimb suspension. J Appl Physiol *73(suppl):* 58S–65S.

Leterme, D., C. Cordonnier, Y. Mounier, M. Falempin (1994) Influence of chronic stretching upon rat soleus muscle during non-weight bearing conditions. Pflügers Arch/Eur J Physiol *429:* 274–279.

Loughna, P., G. Goldspink, D.F. Goldspink (1986) Effect of inactivity and passive stretch on protein turnover in phasic and postural rat muscles. J Appl Physiol *61:* 173–179.

Martin, T.P., A.C. Vailas, J.B. Durivage, V.R. Edgerton, K.R. Castleman (1985) Quantitative histochemical determination of muscle enzymes: biochemical verification. J Histochem Cytochem *33:* 1053–1059.

Morey, E.R. (1979) Spaceflight and bone turnover: Correlation with a new model of weightlessness. Bioscience *29:* 168–172.

Nwoye, L., W.F.H.M. Mommaerts, D.R. Simpson, K. Seraydarian, M. Marusich (1982) Evidence for a direct action of thyroid hormone in specifying muscle properties. Am J Physiol *242:* R401–R408.

Ohira, Y., V.R. Edgerton (1994) Neuromuscular adaptation to gravitational unloading or decreased contractile activity. Adv Exer Sports Physiol *1:* 1–12.

Ohira, Y., B. Jiang, R.R. Roy, V. Oganov, E. Ilyina-Kakueva, J.F. Marini, V.R. Edgerton (1992a) Rat soleus muscle fiber responses to 14 days of spaceflight and hindlimb suspension. J Appl Physiol *73(suppl):* 51S–57S.

Ohira, Y., I. Tabata, H. Shibayama, M. Ohira (1987) Effects of head-down tilt suspension on mass and enzymatic profiles in various types of muscles; in Watanabe S, G Mitdarai, S Mori (eds): Biological Sciences in Space 1986. Tokyo, MYU Research, pp 129–134.

Ohira, Y., T. Wakatsuki, W. Yasui, S. Sugawara, K. Koyanagi, N. Kobayashi (1992b) Effects of tension production and/or neural activity on the regulation of muscle mass during hindlimb suspension in rats; in Matsui N, H Seo (eds): Proc 3rd Int Symp Space Med Nagoya, pp 299–305.

Ohira, Y., W. Yasui, F. Kariya, K. Kaihatsu, K. Nakamura, T. Asakura (1993) Responses of ankle extensor and flexor to immobilization at either stretched or shortened position in rats. Proc 10th Space Utiliz Symp, pp 127–130.

Roy, R.R., K.M. Baldwin, V.R. Edgerton (1991) The plasticity of skeletal muscle: Effects of neuromuscular activity; in Holloszy J (ed): Exercise and Sports Sciences Reviews. Baltimore, Williams & Wilkins, vol 19, pp 269–312.

Roy, R.R., K.M. Baldwin, V.R. Edgerton (1996a) Response of the neuromuscular unit to spaceflight: what has been learned from the rat model?; in Holloszy J (ed): Exercise and Sports Sciences Reviews. Baltimore, Williams & Wilkins, vol 24, pp 399–425.

Roy, R.R., S.C. Graham, V.R. Edgerton (1996b) Architectural properties of chronically unloaded muscles. J Aerospace Env Med *33:* 209–214.

Spector, S.A., C.P. Simard, M. Fournier, E. Sternlicht, V.R. Edgerton (1982) Architectural alterations of rat hind-limb skeletal muscles immobilized at different lengths. Exp Neurol *76:* 94–110.

Thomason, D.B., R.B. Biggs, F.W. Booth (1989) Altered protein metabolism and unchanged β-myosin heavy chain mRNA in unweighted soleus muscle. Am J Physiol *257:* R300–R305.

Thomason, D.B., F.W. Booth (1990) Atrophy of the soleus muscle by hindlimb unweighting. J Appl Physiol *68:* 1–12.

Thomason, D.B., R.E. Herrick, D. Surdyka, K.M. Baldwin (1987) Time course of soleus muscle myosin expression during hindlimb suspension and recovery. J Appl Physiol *63:* 130–137.

Weisberg, A., S. Winegrad, M. Tucker, G. McClellan (1982) Histochemical detection of specific isozymes of myosin in rat ventricular cells. Circ Res *51:* 802–809.

Acta Anatomica

Acta Anat 1997;159:99–107

T. Hijikata
H. Ishikawa

Department of Anatomy,
Gunma University School of Medicine,
Maebashi, Gunma, Japan

Functional Morphology of Serially Linked Skeletal Muscle Fibers

Key Words
Muscle
Tendinous intersection
Nonspanning fibers
Myomuscular junction
Force transmission

Abstract
In the skeletal muscle fiber organization of many vertebrate muscles, serial arrangements or linkages of muscle fibers along the muscle or fascicle are commonly found. These serially linked muscle fibers employ distinct junctional morphologies from muscle to muscle. Notable are the end-to-end linkages of muscle fibers through tendinous intersections (TIs), where many fibers end onto a continuous connective tissue plate with folded terminations similar to myotendinous junctions. Besides this end-to-end linkage, overlapping linkages or arrangements occur among nonspanning fibers terminating intrafascicularly. These nonspanning fibers bear tapering terminations with direct cell-cell (myomuscular) junctions or without any specialized junctions. Despite their overlapping linkages or tapering profiles, nonspanning fibers maintain a uniform sarcomere length along the linked fibers, suggesting that the overlapping-linked nonspanning fibers are equivalent to the end-to-end linked fibers in their mechanical capacity. However, the junctional compliance could differ in their extracellular elastic components and their organization at junctional sites, e.g., direct mechanical (myomuscular) junctions vs. indirect linkages through connective tissue. Increasing evidence suggests that the elastic components, including muscle fibers as well as connective tissues, are more critical than previously thought for the mode and/or the efficiency of tension transmission among serially arranged fibers and thus for the mechanical properties of the muscle.

Introduction

It is still almost generally accepted that individual muscle fibers run the entire length of the fascicles from the origin to the insertion. For example, anatomical and histological textbooks describe that in muscles that do not taper at the ends, such as the sartorius, the fibers apparently continue without interruption through the entire length of the muscle. Contradictory to this concept, however, numerous studies have documented in various vertebrate skeletal muscles including the sartorius, that muscle fibers terminate intrafascicularly [Bardeen, 1903; Huber, 1916; Adrian, 1925; Barrett, 1962; Richmond and Armstrong, 1988; Ounjian et al., 1991; for review see Trotter, 1993]. From these findings, it is now evolving that the serial arrangement of short nonspanning fibers terminating intrafascicularly is a common pattern within vertebrate muscles [Gans, 1989]. At the same time, the presence of a serial arrangement of nonspanning fibers raises some important questions as to muscle mechanics, including the mechanism of force transmission as well as motor control. This paper will survey the intramuscular organization of muscle fibers, the morpho-

KARGER

Fax +41 61 306 12 34
E-Mail karger@karger.ch
www.karger.com

0001–5180/97/1593–0099$15.00/0

This article is also accessible online at:
http://BioMedNet.com/karger

Dr. Takao Hijikata
Department of Anatomy
Gunma University School of Medicine
3-39-22 Showamachi, Maebashi, Gunma 371 (Japan)
Tel. +81 272 20 7912, Fax +81 272 20 7916

logical features of serially linked muscle fibers, including their junctional sites, and discuss their functional and biological implications.

Intramuscular Organization of Skeletal Muscle Fibers: The Occurrence of Connections between Muscle Fibers

The intramuscular organization of skeletal muscle fibers can be classified into three types (fig. 1). One is a simple organization that individual muscle fibers run the entire length of each fascicle from the origin to the insertion (fig. 1a). Another is found in some muscles with tendinous intersections (TIs), e.g., the rectus abdominis, semitendinosus and splenius, where muscle fibers are linked end-to-end through the TI (fig. 1b). More complicated organizations are often found in long parallel-fibered muscles. Some individual fibers run only part of the fascicle length and terminate intrafascicularly (fig. 1c). These short nonspanning fibers are serially linked in overlapping arrays, forming a long fascicle from the origin to the insertion. Each skeletal muscle does not always represent a single type of intramuscular organization. For example, the cat splenius muscle is composed of a combination of all three [Richmond et al., 1985]. On the other hand, even homologous muscles, e.g., the diaphragm [Gordon et al., 1989] and splenius [Richmond et al., 1985; Pfister and Zenker, 1984], have different muscle fiber organizations among mammalian species.

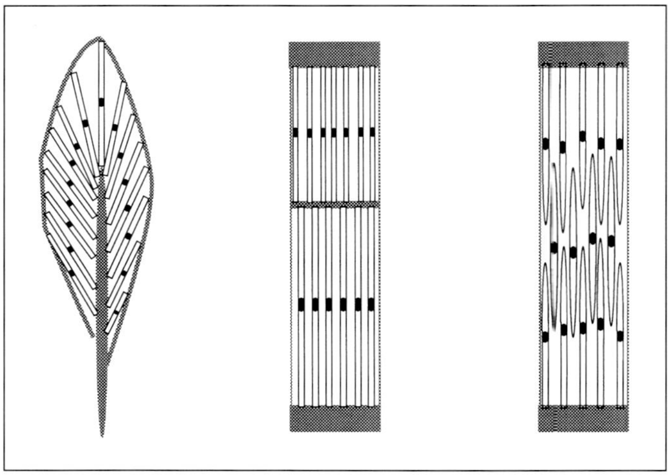

Fig. 1. Schematic drawings of three types of muscle fiber organizations. Gray areas or bands represent tendons or tendinous intersections, while black dots within the fibers indicate motor endplates. See text for explanation.

Many studies dealing with muscle fiber organization have used acetylcholinesterase (AchE) staining to detect motor endplates, based on the assumption that most muscle fibers are singly and focally innervated. This staining is a good method to visualize the muscle fiber organization in each skeletal muscle. Rat skeletal muscles representing each type of fiber organization shown in figure 1 are stained using AchE and are shown in figure 2. The gastrocnemius is a pinnate muscle in which muscle fascicles run obliquely between two or more planes of attachment (fig. 2a). It has a single AchE-positive band along the middle portion of the muscle fascicles, indicating that individual fibers span between the planes of attachment. In contrast, the rectus abdominis muscle, a multibellied muscle, exhibits several endplate bands [Hijikata et al., 1992]. Each band is located at the middle portion between adjacent TIs which are also positive for AchE activity (fig. 2b). Here, most of the muscle fibers span between adjacent TIs, and are linked end-to-end through TIs. Topologically, however, nonspanning fibers terminating intrafascicularly also occur at the lateral and/or medial side of the muscle where the incompletely intersecting TIs permit many fascicles to run interrupted. The frequency of nonspanning fibers varies from fascicle to fascicle (5–100%). The gracilis anterior muscle is a strap muscle without TIs, but possesses double bands of motor endplates along each fascicle (fig. 2c). Microdissection of muscle fibers has revealed that some muscle fibers span between the origin and the insertion, but most fibers (more than 80%) are attached to either the origin or the insertion at one end, but terminate intrafascicularly at the other end [Wines and Hall-Craggs, 1986; Zenker et al., 1990]. In rat this type of organization is observed in the latissimus dorsi and transverse abdominis as well.

Considerations for the Principle that Governs Muscle Architecture with TIs or Nonspanning Fibers

The muscle architecture involving TIs appear to reflect phylogeny and/or ontogeny, since homologous muscles in various mammalian species always possess TIs. However, the detailed architecture of muscles with TIs seem to be more or less modified according to the demands of functions varying across species or topologically [Hijikata et al., 1992]. For example, the mammalian rectus abdominis muscle (RA) is segmented by TIs into serially arranged compartments which are supplied by different thoracic nerves. This segmentation by TIs and separate nerve supplies to compartments might originate phylogenetically from trunk

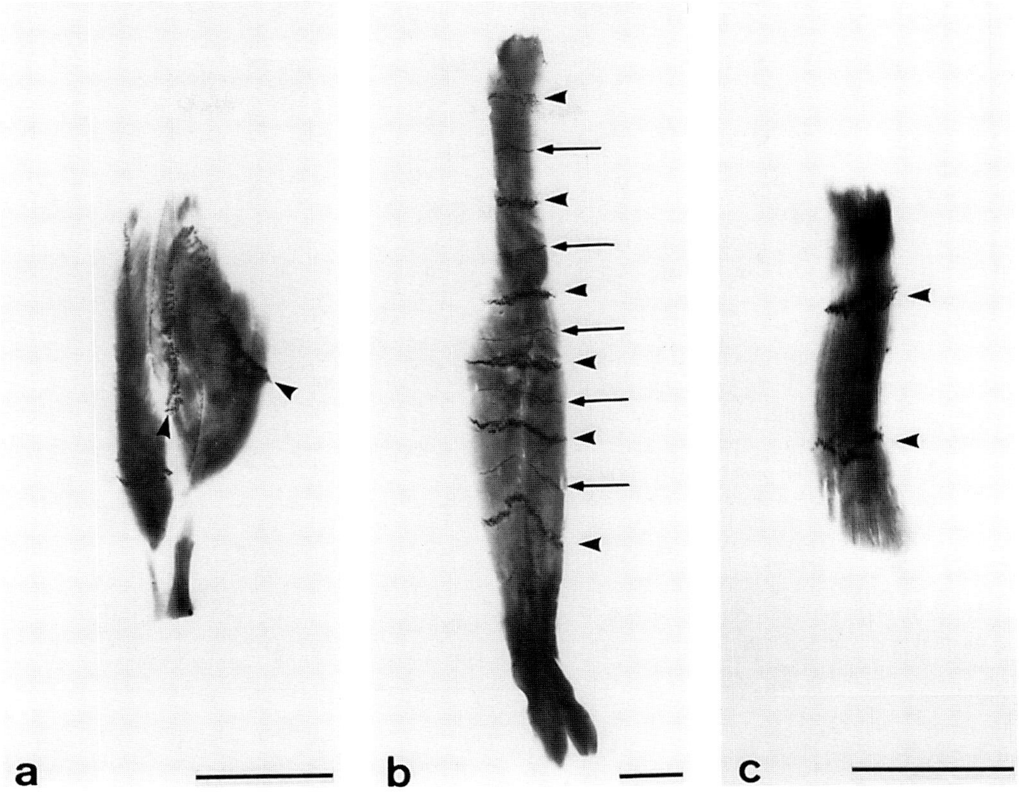

Fig. 2. Photographs of the rat gastrocnemius (**a**), rectus abdominis (**b**), and gracilis anterior muscles (**c**) stained for AchE activity. In addition to broad bands of motor endplates (arrowheads), tendinous intersections are also stained as thinner AchE-positive bands (arrows). See text for explanation. Bars = 1 cm.

muscles of fish or ontogenetically from myomeres of embryonic somites [Ruge, 1892; Braus, 1954; Heinze, 1970]. As shown in table 1, however, each compartment in the RA is not always supplied by a single thoracic nerve and variations in the number of TIs among mammals do not follow a phylogenetic relationship [Eisler, 1912; Konno and Wajima, 1955; Heinze, 1971d]. The supply of multiple nerves to a single cranial and/or caudal segment of RA, probably resulting from the disappearance of TIs, would be reflected from topological and interspecies differences in function.

A plausible explanation has been provided for the muscle architecture with short nonspanning fibers [Loeb et al., 1987]. Short nonspanning fibers and their serial arrangement in long muscles have been regarded as an important means of ensuring mechanically stable contractions of individual muscle fibers. Because conduction velocities in mammalian muscle fibers are slow, typically 2–10 m/s [Schwartz-Giblin et al., 1984], muscle fibers spanning long lengths would have propagation times in excess of their twitch rise times and could not contract simultaneously at all sites along the muscle fiber. Indeed, some limits to fiber length, around 2–3 cm, have been noted for long muscles of relatively large experimental animals [Adrian, 1925; Richmond et al., 1985; Gans et al., 1989; Chanaud et al., 1991]. However, the muscle or fascicle length does not necessarily relate to the occurrence and/or the frequency of short nonspanning fibers [Gordon et al., 1989; Gaunt and Gans, 1990; Ounjian et al., 1991; Hijikata et al., 1992]. The occurrence of short nonspanning fibers or fiber lengths varies among muscles within as well as across species.

The Functional Consequences of Tapering Profiles and Overlapping Arrangements in Nonspanning Fibers

Nonspanning fibers of many in-series muscles exhibit two common features as follows [Richmond et al., 1985; Loeb et al., 1987; Gans et al., 1989; Chanaud et al., 1991;

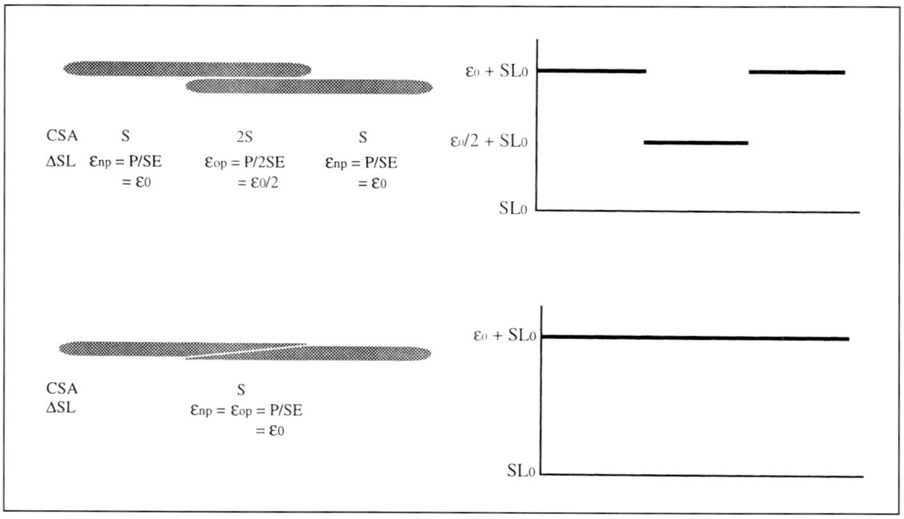

Fig. 3. Schematic diagrams of the overlapping linkages of non-tapering and tapering fibers. In the overlapping linkages of non-tapering fibers with similar CSA (S) along their entire lengths, the strain (ΔSL) differs between the overlapping (ε_{op}) and nonoverlapping portions (ε_{np}). Thus the sarcomere lengths also differ between these regions. P represents the stretching load, E the modulus of longitudinal elasticity, and SL_0 the original sarcomere length. In contrast, the strain and the sarcomere lengths are constant along the linked tapering fibers which adjust their total CSA of overlapping portions to be equal to the CSA in the nonoverlapping portions.

Table 1. The number of compartments segmented by TIs and innervating thoracic nerves in mammalian rectus abdominis muscles

Order Species or common name	Number of compartments	Number of thoracic nerves[1]	Literature
Rodentia			
Rat	7–8	10–11	Heinze [1971b] Hijikata et al. [1992]
Nutria	8–9	13	Heinze [1971b]
Lagomorpha			
Rabbit	10–11	14	Heinze [1971b]
Carnivora			
Cat	7–9	13	Heinze [1971b]
Dog	6–8	9–11	Heinze [1971d]
Perisodactyla			
Horse	11	13	Heinze [1971b]
Artiodactyla			
Domestic cattle	7–8	10	Heinze [1971c]
Domestic sheep	7–8	9–10	Heinze [1971c]
Domestic pig	11–13	13	Heinze [1971a]
Primate			
Macacus cyclopsis	7–9	10–12	Matsuo [1964]
Human	3–6	6–8	Konno and Wajima [1955]

[1] The number of thoracic nerves is more than that of compartments segmentated by TIs. The supply of multiple nerve to a single compartment tends to occur in the cranial and caudal compartments.

Trotter, 1991; Trotter et al., 1992]: (1) they overlap each other for long distances, and (2) they represent a tapering profile (a narrowing caliber diameter) at their intrafascicular termination sites. This tapering profile would be indispensable for ensuring the overlapping linkage of nonspanning fibers to maintain a uniform sarcomere length within a single fiber. If an overlapping linkage of muscle fibers without tapering profiles existed, it would cause mechanical instability within a single fiber. The overlapping linkage of two fibers with a similar cross-sectional area (CSA) along their entire lengths would result in different strains between overlapping and nonoverlapping portions during muscle stretching because the strain is inversely proportional to the CSA (fig. 3). Accordingly, the sarcomere length also would differ between the two portions. Since length-tension curves for vertebrate sarcomeres show an inverted-U shape [Edman, 1966; Gordon et al., 1966; Edman and Reggiani, 1987], these different sarcomere lengths would generate unequal forces within each fiber, resulting in a disrupted contraction. To circumvent this problem, tapering profiles of nonspanning fibers, where the CSA progressively decreases, likely accompany their overlapping linkages.

In support of this hypothesis, morphometric analyses of nonspanning fibers dissected from rat rectus abdominis muscles have indicated that the tapering segments are generally the overlapping portions of the fibers [Hijikata et al., 1993]. The nonspanning fibers overlap over approximately 41–49% of their entire length, while the length of the tapering segments is estimated to be approximately 42% of the entire length of nonspanning fibers. In addition, despite their overlapping linkages, nonspanning fibers maintain a fairly uniform length irrespective of their overlapping and nonoverlapping portions. Nonspanning fibers are considered to adjust the total CSA in the overlapping portion of linked fibers to be nearly equal to the CSA in the nonoverlapping portion of each fiber. In this regard, Bormioli and Schiaffino [1977] have traced nonspanning fibers in serial cross-sections, and have reported that the summated CSA of linked fibers at junctional sites is equal to the CSA of the single fiber beyond the junctional sites.

How do serially linked nonspanning fibers adjust the total CSA in the overlapping portion and how is the differentiation of tapering and nontapering segments produced within nonspanning fibers? This differentiation might result from a mechanosensory mechanism and its feedback signals in muscle fibers. Tapering and non-tapering segments might be produced by different strains caused by overlapping linkages. Larger strains in the non-overlapping portions provide a greater degree of stretching and over the long term may produce hypertrophy of fibers [Lalatta Costerbosa et al., 1987], while smaller strains could possibly lead to an atrophy of the fibers. In this context, it would be interesting to determine whether the reduction ratio of the diameter of nonspanning fibers differs between adult animals and newborn animals which have much smaller strains.

Serially-linked nonspanning fibers have similar CSAs, that is, a similar number of sarcomeres arranged in parallel, along the entire length. Thus, they should have mechanical capacities equivalent to a single fiber or to end-to-end linked fibers with similar CSAs and lengths. However, the stiffness or compliance at the junctional sites must be considered in determining the actual mechanical properties of linked fibers.

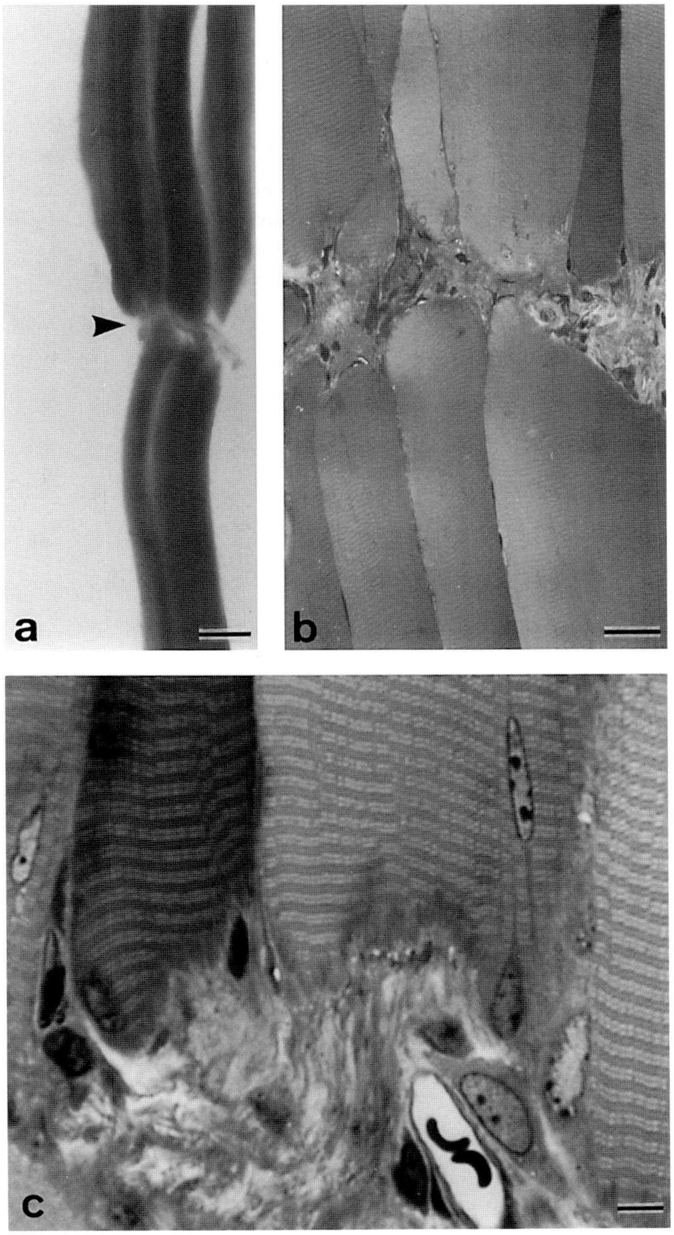

Fig. 4. The termination mode of muscle fibers at a TI in the rat rectus abdominis muscle. **a** Muscle fascicles with TI (arrowhead), stained by gold chrolide. Bar = 100 μm. **b**, **c** Micrographs of the terminal structure of muscle fibers at a TI. Bar = 25 μm.

Junctional Morphology of Serially Linked Muscle Fibers

The Linkages of Muscle Fibers through TIs

Gold chloride staining clearly shows TIs appearing as transparent bands intervening between rounded ends of muscle fibers (fig. 4). Microscopically, the terminal structure of the muscle fibers at the TI is quite similar to that observed at myotendinous junctions (MTJ), referred to as a symmetrical MTJ. The terminal sarcolemma of muscle fibers exhibits many foldings or invaginations parallel to the longitudinal axis of the muscle fibers (fig. 4c). Toward these sarcolemmal invaginations, collagen strands run longitudinally within the TIs. In addition to these longitudinal collagen strands, TIs contain many transverse collagen strands, capillaries, and fibroblasts. Thus, elastic plates link muscle fibers to other muscle fibers at TI-muscle junctions.

The Linkages of Nonspanning Fibers by Myomuscular Junctions or through Endomysium

Nonspanning fibers are linked to each other directly by myomuscular junctions or indirectly via endomysium. In comparison with the indirect connection through the endomysium, the myomuscular junctions appear to be relatively rare or their occurrence may be restricted to small muscles or animals. So far, they have been found in cat and rat extraocular muscles, rat rectus abdominis, gracilis anterior and latissimus dorsi [Teräväinen, 1969; Floyd, 1970; Mayr et al., 1975; Torigoe and Nakamura, 1987; Zenker et al., 1990; Hijikata et al., 1993]. Another difference between these types of linkages is found in the AchE activities. The myomuscular junctions are stained by AchE (fig. 5a), whereas the indirect connections through the endomysium are not.

Myomuscular junctions occur as side-to-side or end-to-side mechanical junctions or as a combination of both [Mayr et al., 1967; Zenker et al., 1990] (fig. 5b). Studies with the transmission electron microscope have shown that myomuscular junctions are interdigitated structures of invaginated sarcolemma [Teräväinen, 1969; Hijikata et al., 1993]. In contrast to myotendinous junctions in TIs and MTJs, this junction type is almost devoid of collagen fibrils between apposed sarcolemmae of linked fibers. Apposed sarcolemmae share a single lamina densa of the basement membrane in some regions, yet have their own lamina densa in other regions (fig. 5c). Apparently, the myomuscular junction has no nexus-like or gap junction-like structure, suggesting that action potentials could not propagate beyond the myomuscular junctions. On the sarcoplasmic side, the sarcolemma is covered with electron-dense materials, into which are inserted thin filaments of terminal sarcomeres.

Trotter [1990] reported the presence of another type of connection between nonspanning fibers in cat biceps femoris and tenuissimus muscles, where nonspanning fibers have insignificant numbers of either myomuscular junctions or specialized connective tissue junctions. Rather, the surface of the tapering portion of the fiber is generally smooth and covered by a continuous external lamina, and varying amounts of collagen endomysium separate the tapered end from adjacent fibers. This arrangement might be considered an indirect connection through the endomysium.

Junctional Compliance in Serially Linked Muscle Fibers

In serially linked muscle fibers, tension generated by myofilaments of a fiber is transmitted to the sarcolemma, the extracellular matrix to include the basement membranes, and to the sarcolemma of linked fibers. Thus it is apparent that the transmission efficiency of tension is influenced by the extracellular compliance at the junctional sites. As described above, the extracellular components and their organization differ among serially linked fibers, and thus these differences would reflect on the mode and efficiency of tension transmission.

The linkages of muscle fibers through TIs would have considerable extracellular stiffness, since a TI is a continuous elastic plate composed of connective tissue, capillaries, and fibroblasts. When motor units are activated during contraction, the exerted force would pull the restricted part of the TI into which activated muscle fibers are inserted, and provide a distributed pull on all of the muscle fibers inserting on the opposite side of the TI. In this respect, these muscle fibers would be serving as an elastic component in the force generation. This stiffness-compliance property of the muscle-connective tissue unit has been observed at the whole muscle level when activating the proximal and distal compartments of the cat semitendinosus muscle independently [Bodine et al., 1982].

In contrast, myomuscular junctions would provide more selective tension to linked fibers. In addition, they would have little extracellular compliance between linked fibers. The folded morphology of sarcolemma at myomuscular junctions increases the surface area and reduces the stress at the interface, as has been suggested for MTJs [Trotter et al., 1981; Tidball, 1984]. Apposed sarcolemmae at myomuscular junctions are closely connected to each other via one or two lamina densa. Accordingly, it is reasonable to suppose that myomuscular junctions are an effective structure for transmitting tension between muscle fibers.

In the indirect connection of nonspanning fibers through endomysium, the concept of lateral transmission has been reassumed [Street and Ramsey, 1965; Street, 1983; Trotter, 1990; Trotter and Purslow, 1992; for review see Trotter, 1993]. In brief, the transmission of tension occurs not only

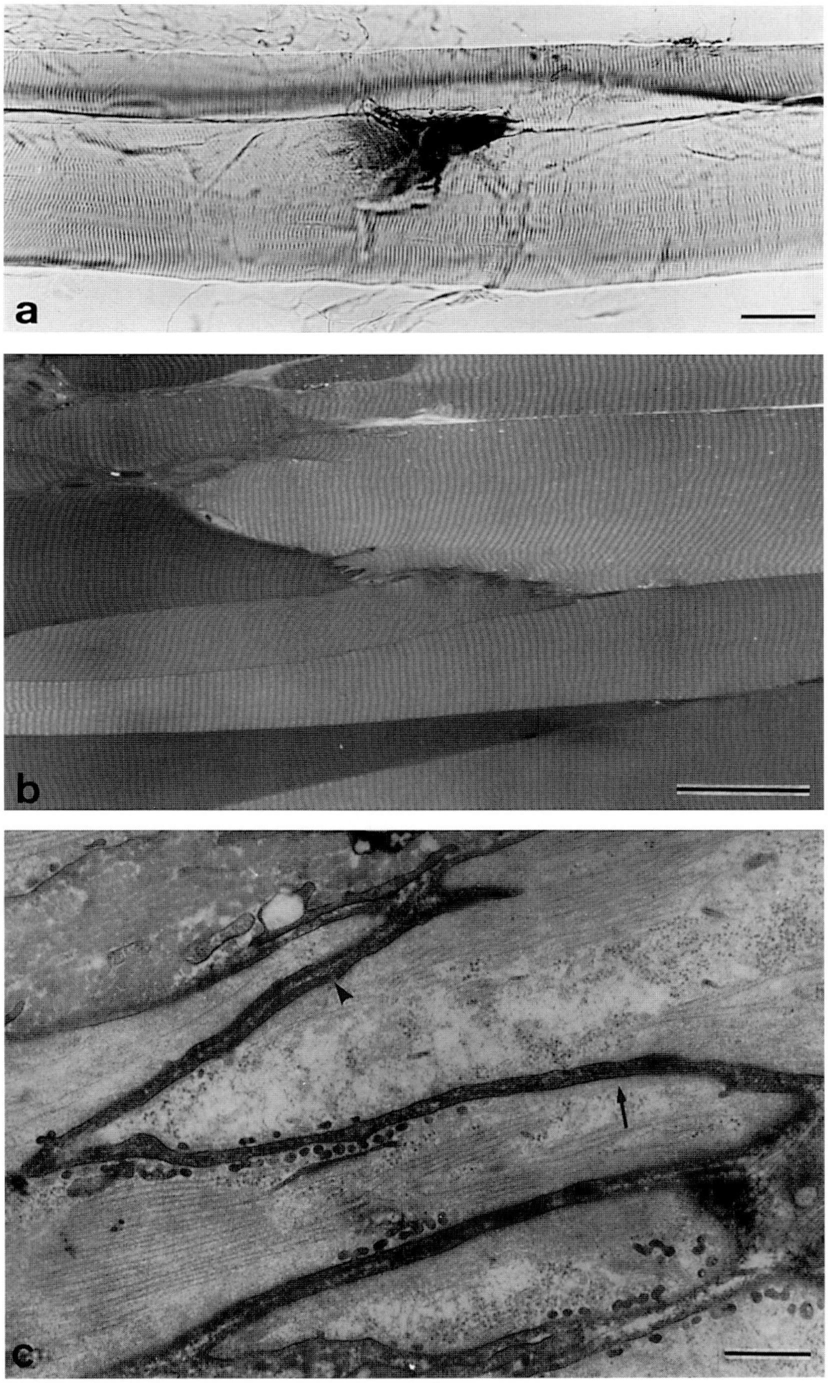

Fig. 5. Myomuscular junctions observed in the rat rectus abdominis muscle. AchE activity is found at the myomuscular junctions occurring as side-to-side connections (**a**). Myomuscular junctions exhibit finger-like processes and invaginations of sarcolemma (**b**, **c**). Apposed sarcolemmae share a single lamina densa (arrow in **c**) in some regions, but have their own lamina densa (arrowhead) in other regions [Hijikata et al., 1993]. Bars = 50 μm in **a** and **b**, 0.5 μm in **c**.

at the tapering ends of the nonspanning fibers, but also in their non-tapered regions. The endomysium surrounding the entire nonspanning fiber likely transfers stress by trans-endomysial shear forces. This concept, on the other hand, means that tension produced by a nonspanning fiber is transferred to all adjacent fibers, with inactive fibers adding extracellular compliance. In this transmission mode, trans-endomysial shear forces seem to permit more or less sliding between adjacent fibers during extreme stretch of the muscle or the endomysium itself might have a buffering effect on the tension. A small discrepancy between changes in fascicle length and sarcomere length has been noted in the cat biceps femoris composed of serially linked fibers [Chanaud et al., 1991].

Concluding Remarks

Serial linkages or arrangements of skeletal muscle fibers occur more commonly than has usually been thought. Serially linked fibers represent different junctional morphologies, especially in the extracellular components and their organization. These differences could reflect on the mode and/or efficiency of tension transmission. In addition to the extracellular connective tissue, each fiber of serially linked fibers can also add in series elasticity if inactive. Thus, the temporal sequence of activation of serially linked fibers could be a critical factor in determining the mechanical properties of a muscle conjoint of serially linked fibers.

Besides the effects on muscle mechanics, the muscle fiber organization might account for some phenomena associated with muscle pathogenesis. Duchenne muscular dystrophy (DMD), the most common of the human muscular dystrophies, is caused by the lack of dystrophin [Hoffman et al., 1987]. Dystrophin is located at the cytoplasmic surface membrane, including the MTJ [Watkins et al., 1988; Masuda et al., 1992], and is assumed to be a structural component to stabilize the sarcolemma [Menke and Jockusch, 1991; Petrof et al., 1993]. In mdx (X-linked muscular dystrophy) mice lacking dystrophin, only the diaphragm displays the muscle pathology typical of human DMD patients [Stedman et al., 1991]. This phenomenon might be explained by different muscle fiber organizations between humans and mice. In the case of the in-series muscle, the damage of even a single muscle fiber could lead to the degeneration of many muscle fibers linked to the initially damaged fibers. In the arrangement involving lateral transmission through the endomysium, nonactivated fibers also would be constantly exposed to stress. This stress could work as a factor to cause the degeneration of fragile muscle fibers lacking dystrophin [Petrof et al., 1993]. Therefore, attention should be given to the type of fiber organization which may be the most susceptible to degeneration in DMD.

References

Adrian, E.D. (1925) The spread of activity in the tenuissimus muscle of the cat and in other complex muscles. J Physiol (Lond) *60:* 301–315.

Bardeen, C.R. (1903) Variations in the internal architecture of the M. obliquus abdominis externus in certain animals. Anat Anz *23:* 241–249.

Barrett, B. (1962) The length and mode of termination of individual muscle fibers in the human sartorius and posterior femoral muscles. Acta Anat *48:* 242–257.

Bodine, S.C., R.R. Roy, D.A. Meadows, R.F. Zernicke, R.D. Sacks, M. Fournier, V.R. Edgerton (1982) Architectural, histochemical and contractile characteristics of a unique biarticular muscle : The cat semitendinosus. J Neurophysiol *48:* 192–201.

Bormioli, S.P., S. Schiaffino (1977) Myomuscular junctions in re-innervated rat skeletal muscle. J Anat *124:* 359–370.

Braus, H. (1954) Anatomie des Menschen. Berlin, Springer, Bd 1, Bewegungsapparat, pp 1–162.

Chanaud, C.M., C.A. Pratt, G.E. Loeb (1991) Functionally complex muscles of the cat hindlimb. II. Mechanical and architectural heterogeneity within the biceps femoris. Exp Brain Res *85:* 257–270.

Edman, K.A.P. (1966) The relation between sarcomere length and active tension in isolated semitendinous fibers of the frog. J Physiol *183:* 407–417.

Edman, K.A.P., C. Reggiani (1987) The sarcomere length-tension relation determined in short segments of intact muscle fibres of the frog. J Physiol *385:* 709–732.

Eisler, P. (1912) Die Muskeln des Stammes; in: Handbuch der Anatomie des Menschen. Jena, Bd II, pp 563–571.

Floyd, K. (1970) Junctions between muscle fibers in cat extraocular muscles. Nature *227:* 185–186.

Gans, C. (1989) Considering functional compartments in muscle. Behav Brain Sci *12:* 654–655.

Gans, C., G.E. Loeb, F. de Vree (1989) Architecture and consequent physiological properties of the semitendinosus muscle in domestic goats. J Morphol *199:* 287–297.

Gaunt, A.S., C. Gans (1990) Architecture of chicken muscles: Short-fibre patterns and their ontogeny. Proc R Soc Lond [Biol] *240:* 351–362.

Gordon, D.C., C.G.M. Hammond, J.T. Fisher, F.J.R. Richmond (1989) Muscle-fiber architecture, innervation, and histochemistry in the diaphragm of the cat. J Morphol *201:* 131–143.

Gordon, A.M., A.F. Huxley, F.J. Julian (1966) The variation in isometric tension with sarcomere length in vertebrate muscles fibers. J Physiol *184:* 170–192.

Heinze, W. (1970) Untersuchungen über die Variabilität der Innervation der Skelettmuskulatur am M. triceps surae, M. flexor digitalis pedis superficialis, M. rectus abdominis, am Plexus lumbosacralis and N. phrenicus von einigen Haussäugetieren. Anat Anz *125:* 81–87.

Heinze, W. (1971a) Die Morphologie der Bauchmuskulatur unserer Haussäugetiere, 1. Mitteilung: Problemstellung, Literaturübersicht, Material und Methodik. Die Bauchmuskulatur des Schweines. Anat Anz *129:* 84–104.

Heinze, W. (1971b) Die Morphologie der Bauchmuskeln unserer Haussäugetiere, 2. Mitteilung: Die Bauchmuskeln des Pferdes, Kaninchens, Sumpfbibers, der Katze und der Ratte. Anat Anz *129:* 133–148.

Heinze, W. (1971c) Die Morphologie der Bauchmuskeln unserer Haussäugetiere, 3. Mitteilung: Die Bauchmuskeln des Schafes, des Rindes und der Ziege. Anat Anz *129:* 276–288.

Heinze, W. (1971d) Die Morphologie der Bauchmuskeln unserer Haussäugetiere, 4. Mitteilung: Die Bauchmuskeln des Hundes, Diskussion (zur Frage der Phylogenese und der Funktion des M. rectus abdominis, die funktionelle Bedeutung der Rektusscheide, die Beziehungen zwischen der Dicke, Länge und Ernährung der Muskelfasern), Zusammenfassung, Literaturverzeichnis. Anat Anz *129:* 374–400.

Hijikata, T., H. Wakisaka, S. Niida (1993) Functional combination of tapering profiles and overlapping arrangements in nonspanning skeletal muscle fibers terminating intrafascicularly. Anat Rec *236:* 602–610.

Hijikata, T., H. Wakisaka, T. Yohro (1992) Architectural design, fiber-type composition, and innervation of the rat rectus abdominis muscle. Anat Rec *234:* 500–512.

Hoffman, E.P., R.J. Brown, Jr., L.M. Kunkel (1987) Dystrophin: The protein product of the Duchenne muscular dystrophy locus. Cell *51:* 919–928.

Huber, C. (1916) On the form and arrangement in fasciculi of striated voluntary muscle fibers. Anat Rec *11:* 149–168.

Konno, T., K. Wajima (1955) Pri la abdomena rekta muskolo ce japanoj. Nichidai Igaku Zasshi *14:* 1511–1525.

Lalatta Costerbosa, G., A.M. Barazzoni, M.L. Lucchi, R. Bortolami (1987) Histochemical types and sizes of fibers in the rectus abdominis muscle of guinea pig: Adaptive response of pregnancy. Anat Rec *217:* 23–29.

Loeb, G.E., C.A. Pratt, C.M. Chanaud, F.J.R. Richmond (1987) Distribution and innervation of short, interdigitated muscle fibers in parallel-fibered muscles of the cat hindlimb. J Morphol *191:* 1–15.

Masuda, T., N. Fujimaki, E. Ozawa, H. Ishikawa (1992) Confocal laser microscopy of dystrophin localization in guinea pig skeletal muscle fibers. J Cell Biol *119:* 543–548.

Matsuo, T. (1964) The medio-ventral muscles of the abdomen in *Macacus cyclopsis*. Acta Med Nagasaki *8:* 63–80.

Mayr, R., J. Gottschall, H. Gruber, W. Neuhuber (1975) Internal structure of cat extraocular muscle. Anat Embryol *148:* 25–34.

Mayr, R., W. Zenker, H. Gruber (1967) Zwischensehnenfreie Skeletmuskelfaser-Verbindungen. Zellforsch *79:* 319–325.

Menke, A., H. Jockusch (1991) Decreased osmotic stability of dystrophin-less muscle cells from the mdx mouse. Nature *349:* 69–71.

Ounjian, M., R.R. Roy, E. Eldred, A. Garfinkel, J.R. Payne, A. Armstrong, A.W. Toga, V.R. Edgerton (1991) Physiological and developmental implications of motor unit anatomy. J Neurobiol *22:* 547–559.

Petrof, B.J., J.B. Schrager, H.H. Stedman, A.M. Kelly, H.L. Sweeney (1993) Dystrophin protects the sarcolemma from stresses developed during muscle contraction. Proc Natl Acad Sci USA *90:* 3710–3714.

Pfister, J., W. Zenker (1984) The splenius capitis muscle of the rat, architecture and histochemistry, afferent and efferent innervation as compared with that of the quadriceps muscle. Anat Embryol *169:* 79–89.

Richmond, F.J.R., J.B. Armstrong (1988) Fiber architecture and histochemistry in the cat neck muscle, biventer cervicis. J Neurophysiol *60:* 46–59.

Richmond, F.J.R., D.R.R. MacGillis, D.A. Scott (1985) Muscle-fiber compartmentalization in cat splenius muscles. J Neurophysiol *53:* 868–885.

Ruge, G. (1892) Zeugnisse für die metamere Verkürzung des Rumpfes bei Säugetieren; Der Musculus rectus thoraco-abdominalis der Primates: Eine vergleichend-anatomische Untersuchung. Gegenbaurs Morphol Jahrb *19:* 376–427.

Schwarz-Giblin, S., M. Halpern, D.W. Pfaff (1984) Segmental organization of rat longissimus: A muscle involved in lordosis behavior. Brain Res *299:* 247–257.

Stedman, H.H., H.L. Sweeney, J.B. Schrager, H.C. Maguire, R.A. Panettieri, B. Petrof, M. Narusawa, J.M. Leferovich, J.T. Sladky, A.M. Kelly (1991) The mdx mouse diaphragm reproduces the degenerative changes of Duchenne muscular dystrophy. Nature *352:* 536–539.

Street, S.F. (1983) Lateral transmission of tension in frog myofibers: A myofibrillar network and transverse cytoskeletal connections are possible transmitters. J Cell Physiol *114:* 346–364.

Street, S.F., R.W. Ramsey (1965) Sarcolemma: Transmitter of active tension in frog skeletal muscle. Science *149:* 1379–1380.

Teräväinen, H. (1969) Localization of acetylcholinesterase activity in myotendinous and myomyous junctions of the striated skeletal muscle of the rat. Experientia *25:* 524–525.

Tidball, J.G. (1984) Myotendinous junction: Morphological changes and mechanical failure associated with muscle cell atrophy. Exp Mol Pathol *40:* 1–12.

Torigoe, K., T. Nakamura (1987) Fine structue of myomyous junctions in the mouse skeletal muscles. Tissue Cell *19:* 243–250.

Trotter, J.A. (1990) Interfiber tension transmission in series-fibered muscles of the cat hindlimb. J Morphol *206:* 351–361.

Trotter, J.A. (1991) Dynamic shape of tapered skeletal muscle fibers. J Morphol *207:* 211–223.

Trotter, J.A. (1993) Functional morphology of force tansmission in skeletal muscle. Acta Anat *146:* 205–222.

Trotter, J.A., K. Corbett, B.P. Avner (1981) Structure and function of the murine muscle-tendon junction. Anat Rec *201:* 293–302.

Trotter, J.A., P.P. Purslow (1992) Functional morphology of the endomysium in series fibered muscles. J Morphol *212:* 109–122.

Trotter, J.A., J.D. Salgado, R. Ozbaysal, A.S. Gaunt (1992) The composite structure of quail pectoralis muscle. J Morphol *212:* 27–35.

Watkins, S.C., E.P. Hoffman, H.S. Slayter, L.M. Kunkel (1988) Immunoelectron microscopic localization of dystrophin in myofibres. Nature *333:* 863–866.

Wines, M.M., E.C.B. Hall-Craggs (1986) Neuromuscular relationships in a muscle having segregated motor endplate zones. I. Anatomical and physiological considerations. J Comp Neurol *249:* 147–151.

Zenker, W., D. Snobl, R. Boetschi (1990) Multifocal innervation and muscle length. A morphological study on the role of myo-myonal junctions, fiber branching and multiple innervation in muscles of different size and shape. Anat Embryol *182:* 273–283.

Acta Anatomica

Acta Anat 1997;159:108–113

Appearance of Complex Branched Muscle Fibers Is Associated with a Shift to Slow Muscle Characteristics

T. Tamaki[a]
A. Akatsuka[b]
S. Uchiyama[d]
Y. Uchiyama[c]
T. Shiraishi[a]

[a] Department of Physiology, Division of Human Structure and Function,
[b] Laboratory for Structure and Function Research, and
[c] Department of Orthopedics, Tokai University School of Medicine, Bohseidai, Isehara,
[d] Tokai University School of Physical Education, Hiratsuka, Japan

Key Words
Skeletal muscle
Complex branched fiber
Muscle regeneration
Contractile properties
Scanning electron microscopy

Abstract

To examine the effects of numerous complex branched fibers (CBF) on whole muscle contractile properties, we established a model of myopathic muscles containing a large number of CBF by repeated local injection of bupivacaine hydrochloride (Marcaine) into the plantaris (PLA) muscle. Marcaine injections were administered once weekly for 10 weeks into the right PLA muscles of Wistar male rats. The in situ contractile properties of Marcaine-injected PLA muscles (I-PLA) were examined under urethane anesthesia, and compared with the contralateral (control) PLA muscle (C-PLA). Numerical and morphological examination using the modified nitric acid fiber digestion method and scanning electron microscopy showed that Marcaine resulted in an 8-fold increase in the number of branched fibers in the I-PLA muscles and about 70% of these fibers were CBF. The latter were composed of ten or more muscle fibers fused together along with many thin and thick, long and short daughter branches. The time to peak tension of twitch and tetanus, and 1/2 relaxation time were significantly longer in I-PLA muscles, representing a shift to slow muscle characteristics. However, the total area of slow fibers/muscle cross-sectional area was similar in I-PLA and C-PLA muscles. Aggregates of same-type fibers (slow fibers) with small and large diameters were observed, reflecting an expected cross-sectional property of CBF. Our results suggest that the appearance of several CBF in a muscle is associated with a shift towards slower contractile properties in the affected muscle.

Introduction

In a series of studies from our laboratories, we have reported the presence of branched muscle fibers in normal [Tamaki et al., 1992a], hypertrophied [Tamaki et al., 1992b], compensatory hypertrophied [Tamaki et al., 1996], regenerated [Tamaki and Akatsuka, 1994] and diseased

Abbreviations used in this paper:
CBF = Complex branched fiber; CSA = cross-sectional area; MW = muscle weight; PLA = plantaris muscle; C-PLA = contralateral (control) plantaris muscle; I-PLA = marcaine-injected plantaris muscle; 1/2RT = one-half relaxation time; TPT = time to peak tension.

KARGER
Fax +41 61 306 12 34
E-Mail karger@karger.ch
www.karger.com

0001–5180/97/1593–0108$15.00/0

This article is also accessible online at:
http://BioMedNet.com/karger

Tetsuro Tamaki, PhD
Department of Physiology, Division of Human Structure and Function
Tokai University School of Medicine, Bohseidai, Isehara, Kanagawa 259-11 (Japan)
Tel. +81 463 93 1121 ext. 2524, Fax +81 463 95 0961
E-Mail tamaki@is.icc.u-tokai.ac.jp

(mdx mouse) muscles [Tamaki et al., 1993]. There are few branched fibers in normal muscles (about 0.003% of the total fibers) and these always appear in simple shapes such as the letters X and Y [Tamaki et al., 1992a]. Thus, the presence of these few branched fibers should not affect the contractile properties of the muscle. An increase in the number of branched fibers (about 2-fold compared to control) has been observed after prolonged weight-lifting exercise in rats [Tamaki et al., 1992b]. However, the proportion of these fibers remains low (about 0.006%) and the shape of the branched fibers is still simple like those in the control muscles. Thus, the likelihood that this small increase in the number of branched fibers following weight-lifting training would affect the whole muscle contractile properties is very low. In contrast, there is a large increase in the number of branched fibers in diseased and compensatory hypertrophied muscles [Tamaki et al., 1993, 1996]. In addition, the shape of the branched fibers in these muscles is very complex. The morphological features of complex branched fibers (CBF) include fusion of ten or more thick fibers with many thin daughter branches, forming a structure resembling an 'anastomosing syncytial reticulum'. The significance of a large number of CBF on the functional properties of the whole muscle is unknown. The purpose of this study was to determine if the presence of a large number of CBF (induced by repeated intramuscular injections of bupivacaine hydrochloride [Marcaine®]) alters the contractile properties of the rat plantaris (PLA) muscle.

Materials and Methods

Animals

Wistar male rats (SPF/VAF, n = 5) were housed individually in standard cages and provided food and water ad libitum. The animals were kept in a constant temperature (23 ± 1 °C) and a 12:12 h light-dark cycle environment throughout the experiment. The first injection of Marcaine was performed at 14 weeks of age (380–420 g body weight). All experimental procedures were conducted in accordance with the Japanese Physiological Society Guidelines for the Care and Use of Laboratory Animals as approved by the Tokai University School of Medicine Committee on Animal Care and Use.

Experimental Myopathy

Experimental myopathy was produced by repetitive injection of Marcaine as previously described [Tamaki and Akatsuka, 1994]. A solution of Marcaine (0.15 ml of 0.5% w/v in 0.9% physiological saline) was injected directly into the right PLA muscle (I-PLA) through the skin and thin part of the gastrocnemius muscle, once a week for 10 weeks. This procedure results in a degeneration-regeneration state of muscle fibers and the formation of many CBF. The untreated contralateral PLA was used as an internal control muscle (C-PLA). Measurement of the contractile properties was performed 2 weeks after the last injection of Marcaine.

Measurement of Contractile Properties

Measurement of contractile properties of the muscle was performed under urethane anesthesia (800 mg/kg i.p.). The body (rectal) temperature was kept constant (approximately 37 °C) using a heated pad. Atropine sulfate was administered subcutaneously at a dose of 0.05 mg/kg to avoid parasympathetic secretory hyperfunction. The trachea was dissected and a tracheostoma tube was secured in place to allow stable spontaneous respiration. The jugular vein was also dissected and cannulated, and warm Ringer's solution containing 5% glucose, was administered intravenously, when necessary. Next, the animal was transferred to a custom-made operation table that allowed stabilization of the head and limbs in a prone position using surgical tape. A midsagittal incision was made from the popliteal area to the base of the calcaneus to expose the posterior aspect of the lower hindlimb. The PLA muscles of both hindlimbs were exposed and freed from surrounding tissues. A stainless steel hook was tied to the distal tendon of each PLA muscle using a silk ligature, and attached to a force-distance transducer (SB-1T, Nihon Kohden, Tokyo) connected to a carrier amplifier (AP-621G, Nihon Kohden). Care was taken to avoid interference with the normal blood supply of the PLA muscle. The surface of the PLA muscle and other exposed tissues was coated with but not immersed in mineral oil to prevent tissue drying and minimize electric noise. The surrounding skin was used to form a mineral oil bath which was maintained at 35 ± 1.5 °C using radiant heat. The sciatic nerve (about 15 mm long) was carefully dissected and exposed under the gluteus medius muscle and also immersed in a small bath of mineral oil with a protective substance (sealon film, Fuji Photo Film Co., Tokyo). Bipolar hooked silver electrodes were placed under the sciatic nerve. Bipolar silver-silver chloride electrodes (1.0 mm diameter) were placed on the surface of the muscle, and the electrical activity of the PLA muscle (EMG) was recorded from the surface of the muscle. The recorded EMG was used to determine the condition of maximal stimulation and other indices of muscle contractility.

Using 1.0-ms-duration single pulses, we recorded muscle twitches at optimum length and need supramaximal stimulation (ranged 1.2–1.5 V) to produce the highest tension and EMG. All mechanical and electrical measurements were recorded on a Linearcorder (Mark VII, WR3101, Graphtec, Tokyo) and an FM tape data recorder (PC208A, Sony Magnescale Inc., Tokyo). Recorded data were stored on a digital audio tape (DAT, DT120RN), and analyzed with a microcomputer (Macintosh Quadra 840-AV, Apple Computer, Inc., Tokyo) using MacADIOS II and SuperScorp II (GW Instruments, Somerville, Mass., USA). Analog-to-digital sampling rate was set at 10 kHz. Peak twitch tension was determined as the average of 10 twitches. The time to peak tension (TPT) and one-half relaxation time (1/2RT) of twitch tensions were measured. Peak tetanic tension was also measured using stimulation frequencies from 10 to 120 Hz. The TPT of tetanus was also measured.

After measurement of the contractile properties, both PLA muscles were excised and weighed after excess connective tissue was removed and then immediately frozen in isopentane precooled by liquid nitrogen and stored at –80 °C. The muscles of 3 animals were prepared for numerical and three-dimensional analysis, and the muscles from the remaining two animals were prepared for histochemical analysis.

Numerical and Three-Dimensional Analysis

The numbers of total and branched fibers were counted using a modified nitric acid digestion method [Tamaki et al., 1989, 1992a]. For this purpose, the frozen muscle was thawed in distilled water and then treated with 15% nitric acid (Wako Pure Chemical Industry,

Table 1. Final body weight, muscle weight, and contractile properties of twitch and tetanus

Body weight	538±45.4 (n=5)		
	I-PLA	% change from control	C-PLA
Muscle weight	638.8±63.3*	+13	564.3±54.7
Twitch			
Tension, N	0.83±0.06	+19	0.70±0.08
Tension, N/kg MW	1.33±0.08	+6	1.25±0.11
TPT, ms	38.78±3.31**	+8	35.98±3.34
½RT, ms	21.67±3.20*	+20	18.09±2.94
Tetanus			
Tension, N	3.47±0.32*	+30	2.66±0.26
Tension, N/kg MW	5.49±0.28	+16	4.75±0.55
TPT, ms	464.3±34.2**	+63	284.3±82.3

Values are mean±SEM. I-PLA=Marcaine-injected plantaris (PLA) muscle; C-PLA=contralateral PLA muscle; MW=muscle weight; TPT=time to peak tension; ½RT=one-half relaxation time.
* $p<0.05$, ** $p<0.01$, compared with C-PLA muscle.

Osaka) for 2–3 h at room temperature. The muscle was then washed in distilled water and cooled for 60 min at 4 °C. The solution was changed from distilled water to 0.01 *M* phosphate-buffered saline (PBS; pH 7.4) and stored at 4 °C. A few bundles were transferred to a silicone-coated Petri dish containing 0.01 *M* PBS (pH 7.4) at room temperature. All fibers were teased free, and the total number of straight and branched fibers was carefully and precisely counted under a dissection microscope (×10–15 magnification).

Some branched fibers were selected for three-dimensional analysis [Tamaki et al., 1992a, 1993; Tamaki and Akatsuka, 1994]. To remove surface connective tissue surrounding the muscle fibers, isolated branched fibers were transferred to glass sample tubes containing 0.01 *M* PBS (pH 7.4) and placed in a hot ultrasonic wave bath at 45–50 °C for 1–3 min. After bath treatment, the fiber samples were fixed in 2.5% glutaraldehyde/0.05 *M* phosphate buffer (PB; pH 7.4) for 2 h, followed by postfixation with 1% osmium tetroxide/0.05 *M* PBS for 1 h at 4 °C. After fixation, the fiber samples were dehydrated with serial graded ethanol solutions, transferred to three changes of 100% t-butyl alcohol and then treated by the t-butyl alcohol freeze-drying method [Inoue and Osatake, 1988]. After spatter-coating with gold in an ion spatter coater (JEOL JFC-1100), the samples were examined by scanning electron microscopy (JEOL JSM-840A).

Histochemical Analysis
The PLA muscles were divided into six blocks at –20 °C, to allow examination of a large section of the muscle [Tamaki and Uchiyama, 1995; Tamaki et al., 1996]. Serial 10-µm-thick transverse sections were obtained from each block (6–10 sections). The sections were stained for actomyosin adenosine triphosphatase (ATPase) after acid (pH 4.35) preincubation [Guth and Samaha, 1970]. Photographs of 6 blocks of each PLA muscle were taken at the same magnification. The number of slow fibers, total area of slow fibers, mean area of single slow fibers, and cross-sectional area (CSA) of each block were analyzed, with and without the tendon, and calculated using a microcomputer and image-scanner.

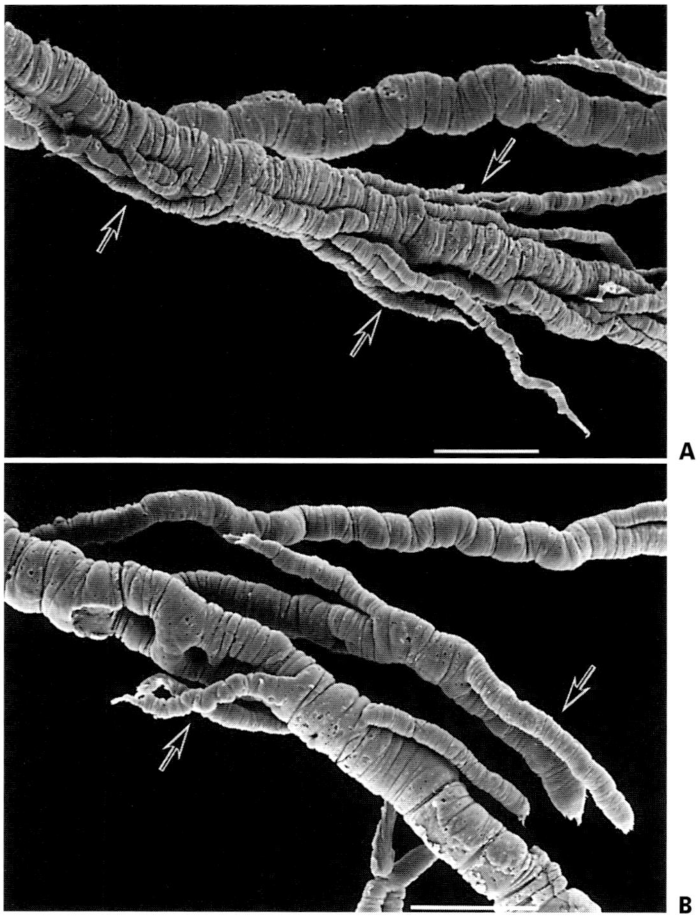

Fig. 1. Typical scanning electron micrograph of complex branched fibers. Note the presence of ten or more fibers fused together with many daughter fibers (**A**, **B**). Note that several thin daughter branches are evident in these complex branched fibers (arrows **A**, **B**). Bar= 100 µm.

Statistical Analysis
Data were expressed as mean ± SEM. Differences between groups were tested with the Student's t test (paired). The level of significance accepted a priori was set at $p<0.05$.

Results

Effect of Marcaine Treatment on Contractile Properties
Final body weight, comparisons of muscle weight, and contractile properties between Marcaine-treated I-PLA and untreated control C-PLA are summarized in table 1. The mean weight of I-PLA muscle was significantly higher

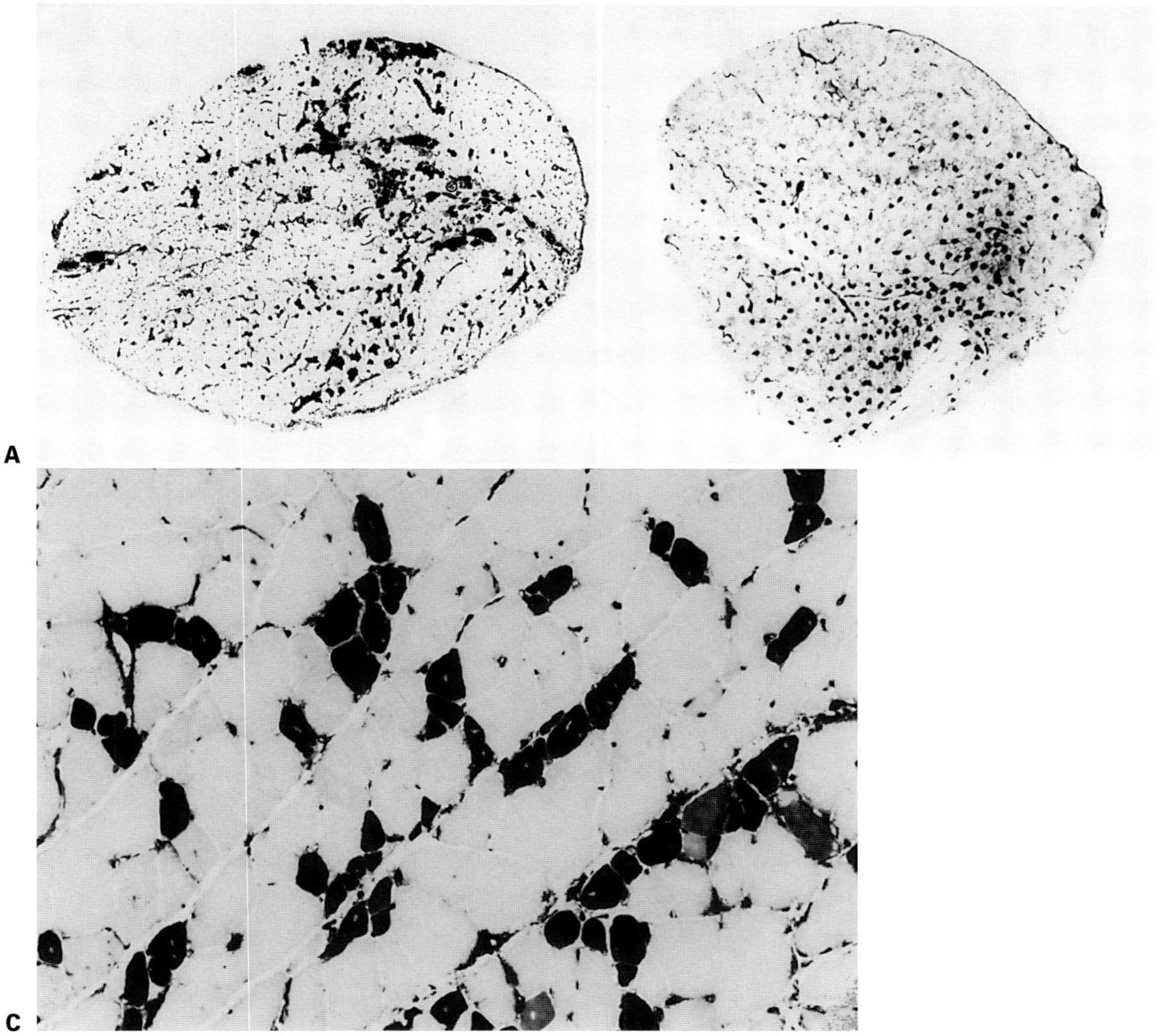

Fig. 2. Histochemical properties of I-PLA (**A**) and C-PLA (**B**) muscles, and high magnification photograph of I-PLA muscle (**C**) stained for ATPase (pH 4.35). Slow fibers (dark staining) sparsely distributed near the tendon within the section of the C-PLA muscle (**B**). Several aggregates of fibers of the same type are evident in the I-PLA muscle (**A**). Note that several fibers with small diameters appeared in these aggregations (**C**), reflecting a typical cross-sectional feature of complex branched fibers as shown in figure 1A, B. Magnification: **A**, **B** ×15; **C** ×188.

(13%) compared with C-PLA muscle. The maximum absolute twitch and tetanic tensions were 19 and 30% higher, respectively, in I-PLA compared with C-PLA muscles, whereas the relative tensions were similar. Both twitch and tetanic TPT and twitch 1/2RT were significantly longer in I-PLA than C-PLA muscles, suggesting a shift to slower muscle characteristics following Marcaine treatment.

Numerical and Three-Dimensional Properties

Treatment with Marcaine resulted in a small, insignificant decrease of total fiber number in I-PLA compared with C-PLA muscles. More importantly, the number of branched fibers was 8-fold higher in I-PLA than C-PLA muscles (table 2). In addition, about 70% of the branched fibers had complex features. Typical features of CBF are shown in figures 1A and B. Marcaine treatment resulted in the fusion of approximately ten or more fibers together and the appearance of many daughter branches.

Histochemical Cross-Sectional Properties

Typical photographs of whole muscle cross-sectional profiles (stained for ATPase, pH 4.35) of the I-PLA and

Table 2. Number of total and branched fibers, and results of analysis of histochemical cross-sections stained for ATPase

	I-PLA	C-PLA
Whole muscular properties (n=3)		
Total fiber number	10,351±294	11,032±126
Number of branched fiber	265±32.1*	33.4±3.2
Cross-sectional properties (n=2)		
Total area of slow fibers, mm^2	0.482±0.23	0.351±0.09
Mean area of single slow fibers, μm^2	1,041±317	1,127±199
Number of slow fibers	463±29**	311±11
Mean CSA, mm^2	12.7±1.1	11.2±0.8
Mean CSA without tendon, mm^2	12.3±0.8	10.9±0.6
Total area of slow fibers/CSA (without tendon) %	3.88±0.9	3.22±0.5

The cross-sectional properties are based on whole muscle cross-sections obtained by uniform divisions of the muscle into six parts along the entire muscle length. Analysis was performed using NIH image analysis software. Values represent mean ± SEM. CSA = Cross-sectional area.

* $p<0.05$, ** $p<0.01$, compared with C-PLA muscle.

C-PLA muscles are shown in figures 2A and B, respectively. These sections were taken from the mid-belly of the distal half of each muscle. The distal tendon is positioned on the right side of each photograph. Dark-stained fibers are slow fibers which were sparsely distributed near the tendon in C-PLA muscle (B). However, aggregations of fibers of the same type were frequently observed in I-PLA muscle (A). Figure 2C shows a higher magnification of fiber aggregations. Aggregations were composed of small and large fibers of the same type representing the cross-sectional profiles of CBF as shown in figure 1. The statistical data from the sections from both I-PLA and C-PLA muscles are summarized in table 2. The number of branched fibers was 8-fold higher in I-PLA than in C-PLA. In addition, the number of slow fibers significantly increased in I-PLA (+49%) compared with C-PLA muscles. The mean CSA, with and without tendon area, increased in I-PLA muscle (+13% and +13%, respectively; $p>0.05$) reflecting the increase in muscle weight.

Discussion

Our results showed that repeated local injections of Marcaine resulted in a significant increase in the weight of I-PLA compared with C-PLA muscles. This may reflect an increase in both the connective tissues and branching of the muscle fibers. The increased muscle weight suggests that considerably large alterations in the muscle were induced by repetitive injection of Marcaine. In addition, an increased tension capability was observed in the I-PLA muscles. The absolute maximum tetanic tension was significantly increased in I-PLA compared to C-PLA muscles (table 1). This increase in tension may be partly related to an increase in the contractile component due to increased branching of the fibers following the formation of CBF. This hypothesis is supported by the 8-fold increase in the number of branched fibers in I-PLA compared with C-PLA muscles (table 2). However, the relative maximum tensions for both the twitch and tetanus were similar in I-PLA and C-PLA muscles. This result suggests that repetitive injection of Marcaine induced an increase in the noncontractile components such as connective tissues, as well as an increase in the CBF within the whole muscle.

The twitch and tetanic TPT and twitch 1/2RT were significantly longer in I-PLA compared with C-PLA muscles (table 1). It appears that I-PLA muscle shifted to slower muscle characteristics. Certainly, the number of slow fibers increased in the I-PLA muscles (+49%). The total area of slow fibers in a cross-section, however, did not significantly increase in I-PLA muscle, although there was a tendency for an increase to occur (+37%). In addition, the mean area of slow fibers tended to decrease in I-PLA muscle (–8%). Several aggregations of slow fibers were evident in I-PLA muscles (fig. 2A). In figure 2C, these aggregations can be seen to include several small fibers and resemble the expected cross-sectional features of CBF as shown in figures 1A and B. Thus, the tendency for an increase in the total area of slow fibers in I-PLA muscle appears to result from the appearance of several aggregations of slow fibers (fig. 2A, C), and the decrease in mean area of single slow fibers appears to result from the increase in slow fibers with small diameters. Thus, our data suggest that the shift to a slower contractility of the I-PLA muscle was associated with the presence of many CBF in the muscle.

We have previously reported that many CBF are also present in compensatory hypertrophied PLA muscles of rats [Tamaki et al., 1996], where a shift towards slower characteristics occurs [Tamaki and Shiraishi, 1996]. These changes have been confirmed by several investigators using this model. For example, slower velocity of shortening [Binkhorst and Van'Thof, 1973; Roy et al., 1982], reduction in the conduction velocity of motoneurons with a reduced cell body size [Finkelstein et al., 1991], reduction in muscle myosin [Baldwin et al., 1982; Nobel et al., 1983] and myofibrillar ATPase activity [Baldwin et al., 1977,

1982; Tsika et al., 1987] have been reported. In our study, the maximum twitch and tetanic tension of the compensatory hypertrophied PLA muscles shifted toward slower characteristics, and these results were reflected in histochemical examination [Tamaki and Shiraishi, 1996]. In particular, a dramatic increase in the number and cross-sectional area of slow fibers was observed in the hypertrophied muscles compared with the contralateral muscles. The percentage of area containing slow fibers within the whole muscle cross-sectional area (excluding the tendon) increased by 4.2% in the overloaded PLA compared with the contralateral PLA muscle. In the fatigue test (twitch train of 15 Hz), however, a significant reduction in fatigue resistance was observed in overloaded PLA muscles. Thus, fatigue resistance of the overloaded PLA was not consistent with the increase in slow, fatigue-resistant fibers. This paradoxical result may be explained by the presence of many CBF which may reduce the fatigue resistance of the whole muscle. A tendency for less fatigue resistance in compensatory hypertrophied muscle has also been reported by Olha et al. [1988].

We have also reported the presence of numerous CBF in the soleus muscles of the mdx mouse [Tamaki et al., 1993]. Carter et al. [1995] reported impaired fatigue resistance during voluntary wheel running and a lesser effect of endurance training on the soleus muscle of the mdx mouse compared with the normal C57BL/10 mouse. This report strongly supports our hypothesis on the relationship between the existence of numerous CBF and the decrease in fatigue resistance.

In conclusion, many CBF are observed in muscles of the mdx mouse, and in compensatory hypertrophied and Marcaine-injected rat PLA muscles. It seems that the presence of a large number of CBF in a muscle shifts the affected muscle to slower contractile properties and reduces fatigue resistance of the whole muscle.

Acknowledgment

We thank Mr. M. Tokunaga, Laboratory for Structure and Function Research, Tokai University School of Medicine, for his technical assistance with photography.

References

Baldwin, K.M., W.G. Cheadle, O.M. Martinez, D.A. Cooke (1977) Effect of functional overload on enzyme levels in different types of skeletal muscle. J Appl Physiol *42:* 312–317.

Baldwin, K.M., V. Valdez, R.E. Herrick, A.M. Macintosh, R.R. Roy (1982) Biochemical properties of overloaded fast-twitch skeletal muscle. J Appl Physiol *52:* 467–472.

Binkhorst, R.A., M.A. Van'Thof (1973) Force-velocity relationship and contraction time of the rat plantaris muscle due to compensatory hypertrophy. Pflügers Arch *342:* 145–158.

Carter, G.T., M.A. Wineinger, S.A. Walsh, S.J. Horasek, R.T. Abresch, W.M. Fowler, Jr. (1995) Effect of voluntary wheel-running exercise on muscles of the mdx mouse. Neuromusc Disord *4:* 323–332.

Finkelstein, D.I., J.G. Lang, A.R. Luff (1991) Functional and structural changes of rat plantaris motoneurons following compensatory hypertrophy of the muscle. Anat Rec *229:* 129–137.

Guth, L., F.J. Samaha (1970) Procedure for the histochemical demonstration of actomyosin ATPase. Exp Neurol *28:* 365–367.

Inoue, T., H. Osatake (1988) A new drying method of biological specimens for scanning electron microscopy: The t-butyl alcohol freeze-drying method. Arch Histol Cytol *51:* 53–59.

Nobel, E.G., B.L. Dabrowski, C.D. Ianuzzo (1983) Myosin transformation in hypertrophied rat muscle. Pflügers Arch *396:* 260–262.

Olha, A.E., B.J. Jasmin, R.N. Michel, P.F. Gardiner (1988) Physiological responses of rat plantaris motor units to overload induced by surgical removal of its synergists. J Neurophysiol *60:* 2138–2151.

Roy, R.R., I.D. Meadows, K.M. Baldwin, V.R. Edgerton (1982) Functional significance of compensatory overloaded rat fast muscle. J Appl Physiol *52:* 473–478.

Tamaki, T., A. Akatsuka (1994) Appearance of complex branched fibers following repetitive muscle trauma in normal rat skeletal muscle. Anat Rec *240:* 217–224.

Tamaki, T., A. Akatsuka, J. Itoh, S. Nakano (1989) A newly modified isolation method of single muscle fibers: Especially useful in histological, histochemical and electron microscopic studies on branched fibers. Tokai J Exp Clin Med *14:* 211–218.

Tamaki, T., A. Akatsuka, M. Tokunaga, S. Uchiyama, T. Shiraishi (1996) Characteristics of compensatory hypertrophied muscle in the rat. I. Electron microscopic and immunohistochemical studies. Anat Rec *246:* 325–334.

Tamaki, T., T. Sekine, A. Akatsuka, S. Uchiyama, S. Nakano (1992a) Detection of neuromuscular junctions on isolated branched muscle fibers: Application of nitric acid digestion method for scanning electron microscopy. J Electron Microsc *41:* 76–81.

Tamaki, T., T. Sekine, A. Akatsuka, S. Uchiyama, S. Nakano (1993) Three-dimensional cytoarchitecture of complex branched fibers in soleus muscle from mdx mutant mice. Anat Rec *237:* 338–344.

Tamaki, T., T. Shiraishi (1996) Characteristics of compensatory hypertrophied muscle in the rat. II. Comparison of histochemical and functional properties. Anat Rec *246:* 335–342.

Tamaki, T., S. Uchiyama (1995) Absolute and relative growth of rat skeletal muscle. Physiol Behav *57:* 913–919.

Tamaki, T., S. Uchiyama, S. Nakano (1992b) A weight-lifting exercise model for inducing hypertrophy in the hindlimb muscles of rats. Med Sci Sports Exerc *24:* 881–886.

Tsika, R.W., R.E. Herrick, K.M. Baldwin (1987) Interaction of compensatory overload and hindlimb suspension on myosin isoform expression. J Appl Physiol *62:* 2180–2186.

Acta Anatomica

Acta Anat 1997;159:114–126

E. Eldred[a, b]
L. Yung[b]
R.R. Roy[b]

[a] Department of Neurobiology, and
[b] Brain Research Institute,
University of California at Los Angeles,
Calif., USA

Spindle Representation relative to Distribution of Muscle Fiber Types in the Cat Capsularis Muscle

Key Words
Cat capsularis muscle
Representation of muscle spindles
Distribution of muscle fiber types

Abstract
The spatial representation of muscle spindles (Sps) in the small (~0.2 g), simply structured capsularis muscle that crosses anterior to the cat's hip joint was compared with the distribution of the slow oxidative (SO) and few (<10%) fast oxidative-glycolytic (FOG) fibers of which it is composed to see if their distributions were consistent with a hypothesis that sensory input from Sps influences the incidence of extrafusal fiber types. In frozen sections from 4 muscles, FOG fibers were enumerated along 1-mm strips across the muscle's maximum width, and between the 'superficial' surface and the 'deep' one that contacts the joint. The locations of Sps in complete serial sections of 2 paraffin-embedded muscles, one perfused with the hip joint flexed and the other with it extended, were plotted on an outline of each muscle at its midlength, and their numbers and density in horizonal and sagittal 'strata' determined. In general, the incidence of Sps increased down the superficial-to-deep axis, while FOG fibers became fewer, as is consistent with support of SO status by Sp input. Along the craniocaudal axis, i.e. width, the numbers of FOG fibers rose toward the hip joint, but this was not associated with a monomodal gradient of Sps. In the extended muscle, however, the lengths of the axial bundle and capsular space of Sps in the half of the muscle next to the joint exceeded those in the longer, cranial half, implying that under stretch the input from Sps became higher toward the joint. In the non-extended muscle these lengths did not differ, although the lengths of extrafusal fibers isolated from 2 macerated muscles and normalized according to sarcomere length decreased linearily by ~50% along the craniocaudal axis. It is explained that if elastic resistance of a Sp's sensory region exceeded that of an equivalent length of septal tissue in-series, the progressive shift in the ratio of compliances across this trapezoidally-shaped muscle should result in relatively greater lengthening of Sps at the shorter border as the muscle was extended. Levels of discharge conducive to transition of some motor units to the FOG type might be attained. Thus, gradients in the discharge of Sps (but not necessarily incidence) along both transverse and superficial-to-deep axes may be consistent with Sp sensory input influencing the distribution of at least these types of motor units.

KARGER

Fax +41 61 306 12 34
E-Mail karger@karger.ch
www.karger.com

This article is also accessible online at:
http://BioMedNet.com/karger

Roland R. Roy, PhD
Brain Research Institute, University of California at Los Angeles
405 Hilgard Avenue, Los Angeles, CA 90095-1761 (USA)
Tel. +1 310 825-2193, Fax +1 310 206-9184
E-Mail rrr@ucla.edu

Introduction

Crossing the cranioinferior aspect of the cat's iliofemoral joint is a small (0.2–0.3 g) muscle, the M. articularis coxae or 'capsularis' which has nearly the highest density of muscle spindles (Sps) reported for a cat limb muscle (156 Sp/g, Eldred, Yung and Roy [unpubl. observation]). About 90% of its fibers are of the slow oxidative (SO) type and the remainder are fast oxidative-glycolytic (FOG). This limitation in fiber type, the high Sp content, and a simple architecture wherein fibers pass from origin to insertion at a low angle (<5°) make the capsularis especially well suited for analysis of relationships between the spatial representation of Sps and distribution of muscle fiber types. In muscles comprised of multiple types of fibers, slow fibers are more common at deep levels (closer to bone) where Sps also are better represented [Botterman et al., 1978; Peck et al., 1984], suggesting that they may have a positive, causative relationship. Slow muscles and slow motor units (MUs) in muscles with multiple fiber types, have strong reflex ties to proprioceptive input [Maier et al., 1972; Peck et al., 1984], and in a variety of experimental procedures shown to alter the slow-fast status of muscles, the input from Sps also is altered. Those that lessen sensory input (table 1, 'reduced input') are followed by a shift of muscles toward a fast status, while procedures predicted to elevate Sp discharge ('increased input, loading') generally are associated with a slowing of contractions and an increase in the representation of slow fibers. Even the fast-to-slow fiber transition in the soleus that appears spontaneously during development is reduced when a rat's leg is casted and presumedly proprioceptive inflow from it is lowered [Ansved, 1995].

In a survey of fiber types in cat thigh muscles [Roy and Acosta, 1986], a SO:FOG fiber ratio of 93:7 was obtained from a sample area near the deep surface of the capsularis, and 76:24 near the superficial surface. The distributions of fiber types through the muscle's thickness and width have been quantified and compared with the Sp representation. Unlike the association of fiber types with the incidence of Sps down the muscle's depth, a shift in fiber type across the muscle was not accompanied by a monomodal gradient of Sps. This leads to consideration of the effects that the trapezoidal shape of this muscle might have on proprioceptive inflow and the SO-FOG status of its MUs.

Materials and Methods

At the conclusion of unrelated experiments, 10 female cats of 3–4 kg weight were terminated with intravenous pentobarbital. In 4 cats one capsularis muscle was removed, mounted upright on cork, quick-frozen in isopentane covered with liquid nitrogen, and 10-μm sections taken at the muscle midlength. Sections were stained for myofibrillar ATPase after alkaline and acid preincubation, succinic dehydrogenase, and α-glycerophosphate dehydrogenase, permitting classification of fibers as FOG or SO [Peter et al., 1972]. In addition, the staining reactions to monoclonal antibodies for specific myosin heavy chain proteins (MHC I, IIa, and IIx) were used to classify the fibers as fast or slow. All fibers that stained dark after alkaline preincubation, i.e. those that were FOG, reacted with monoclonal antibodies BF-35 and BF-71, but not BA-F8, whereas fibers identified as SO reacted with BA-F8 and BF-35 [see Talmadge et al., 1995, for classification scheme]. The location of each FOG fiber on a section at the muscle midlength was plotted on a 100× enlarged sketch, the distinction of fibers in contact with one another being verified under the microscope. Counts were made of the FOG fibers in 0.5×1.0 mm quadrats along 1-mm strips centered over maximal transverse and superficial-to-deep axes of the muscle cross-sectional outline.

Spindles were identified in 2 muscles, one from a cat in which the hindquarters had been perfused with saline followed by buffered formalin while the iliofemoral joint on one side was flexed, and the other from a cat with the thigh held in extension. After resection, the muscles were immersed in Bouin's solution for 72 h, double-embedded in paraffin, and a complete series of 10-μm cross-sections obtained for staining with hematoxylin-and-eosin. The muscle 'at rest length' from the cat with the hip joint flexed yielded 1,270 sections, and the 'extended' one, 2,400, from which 93 and 125 outlines were drawn at 100× on which were entered the locations of Sps and codified notes as to their structure as seen on that section. The locations were then replotted on an outline at the muscle midlength or other levels by tracing each Sp's relation to identified fascicles and tandem combinations. Expression of Sp distribution as units per area at the muscle midlength as done for FOG fibers was not suitable, for Sps were far fewer and cross-sectional areas of the muscle at levels where they were located varied. Instead horizontal 'strata' in thickness a tenth the maximum distance between the muscle's superficial and deep surfaces were outlined (fig. 2A), the Sps in each stratum counted, its area measured, and Sp densities in the respective strata derived. To combine data from the 2 muscles, the means of the counts from corresponding strata were calculated. To combine the expressions of density, Sp counts from the extended muscle were multiplied by the ratio of the total cross-sectional area at midlength of the non-extended to that of the extended muscle, and the mean of this adjusted (increased) value plus that from the shorter muscle were obtained for each stratal level. Ten sagittally oriented strata were similarly delimited (vertical lines in fig. 2A), and for the 2 muscles the means of Sp counts and the adjusted densities were derived.

Lengths of the intrafusal fiber bundle, capsule and intracapsular space were measured in the serially sectioned muscles, every section being examined when a component was about to terminate. In 2 more cats, the lengths of representative extrafusal muscle fibers were obtained. The superficial muscles at one hip joint were removed, the fascial planes between the deeper muscles opened to facilitate fluid entry, and with the hip joint extended, the hindquarters submerged in 10% buffered formalin. After immersion for ~10 days, the capsularis was further exposed, measured in situ, excised without stretching, and then stored in phosphate buffer. The muscle then was macerated in 15% nitric acid and divided into 4 or 5 longitudinal sectors of equal width from which several fibers were isolated under a dissecting microscope. After measuring a fiber's length using vernier calipers, it

Table 1. Long-term procedures that affect the fiber-type and contractile status of a skeletal muscle which also may decrease or increase proprioceptive input[1,2]

Procedure	Effects	Reference
Decreased input		
Deprivation of all afferent input by		
Tetrodotoxin block for 4 weeks; rat S	muscle faster; fiber I:II ratio ↓	Spector [1985]
Deafferentiation at 2–3 days, observed after 12 weeks; cat S, VI, MG, FHL	differentiation of slow muscle ↓ fast muscles no difference	Buller et al. [1960a]
Spinal cord isolation as adult, observed at 6 months; cat, 16 muscles	ratio I:II fibers and fiber CSAs ↓ more than spinalization alone	Roy et al. [1992]
Unloading by		
Hindlimb suspension, 1–4 weeks; rat S	CSA of I ↓ > than II; ratio I:II ↓	Hauschka et al. [1987]
Spaceflight, 6 days; rat S	muscle faster; I:II ratio ↓; CSA slow fibers ↓ more than fast	Caiozzo et al. [1996]
Pinning at rest length, 4 weeks; rat S	muscle faster	Fischbach and Robbins [1969]
Casting at rest length, 4–21 weeks; rat S	muscle faster; I:II ratio ↓	Ansved [1995]
Tenotomy[3], 3–10 weeks; rabbit S, G	fiber atrophy S > G and in G, I > IIb	Engel et al. [1966]
Increased input		
Loading by		
Removal of synergists, 12–14 weeks; rat Pl	muscle slower; less fatigable	Roy et al. [1982]
Removal of synergists, 12–14 weeks; rat Pl	I:II ratio ↑; slow myosin ↑	Baldwin et al. [1982]
Antagonists denervated[3], 8 months; rat S	fiber I:II ratio ↓	Guth and Wells [1972]
Pinning in extension, 4 weeks; rat S	muscle faster	Fischbach and Robbins [1969]
Casting in extension, 3 weeks; rat VI	I:IIa fibers ↑	Boyes and Johnson [1979]
Partial denervation at 3 or 18 days, rat EDL	muscle slower; I:II ↑; fatigue ↓	Tyc and Vrbova [1995]
Postural support after spinalization, 6 months; cat MG, S	S SDH ↑; MG fiber GPD, CSA no difference	Jiang et al. [1990]
After suspension 2 weeks, return to normal activity for 1 week; rat S	return to normal contraction rate and myosin I level	Templeton et al. [1984]
Centrifugation at 2 *g*, 2 weeks; rat S	*** slow myosin ↓	Roy et al. [1996]
Dynamic exercise by[4]		
Swimming, 52 days; rat S, Pl	S, Pl, not altered	Edgerton et al. [1969]
Treadmill after spinalization, 6 months; cat, 16 muscles	* muscle slower; conversion of fibers to fast ↓	Roy and Acosta [1986]
Passive manipulation after spinal cord isolation	** no effect on conversion	Roy et al. [1992]
Treadmill, 9 or 18 weeks; guinea pig MG	I:II fibers ↑ in red and white areas	Barnard et al. [1970]
Treadmill 19% grade during 7 days of suspension; rat MG	** ratio fiber type, SDH and GPD, no difference; fiber CSA ↑	Graham et al. [1989a]
Treadmill 30% grade during 4 weeks, 4-week suspension; rat S	** I:II ratio no difference; fiber CSA, SDH ↑	Graham et al. [1989b]
Treadmill after spinalization at 2 weeks;		
6 or 12 months; cat S, MG	** both muscles no difference in conversion to fast	Baldwin et al. [1984]
6 months; cat S, MG	** contraction time of MUs no difference	Smith et al. [1993]
11–19 weeks; cat S, MG	*** both muscles faster; ratio I:II fibers ↓	Johnson et al. [1982]
Eccentric contractions by 'dropping' during 4 weeks of suspension; rat S	** no difference in conversion to fast	Hauschka et al. [1987]
Climbing with a load during 1 week of suspension; rat S, MG	*** S relaxation time ↓; (CT ↓ at $p<0.10$); MG, no difference	Herbert et al. [1988]

[1] Entries are intended to be illustrative, and do not include all relevant references, e.g. effects of spinalization alone are not included. Where an exercise is imposed in the presence of a background procedure, the 'effect' refers to comparison with and without the exercise.

[2] Abbreviations are for slow muscles, S=soleus, and VI=vastus intermedius; and for primarily fast muscles, FHL=flexor hallucis longus, MG=medial gastrocnemius, Pl=plantaris, TA=tibialis anterior, VL=vastus lateralis; and for fibers, SDH=succinate dehydrogenase; GPD=α-glycerophosphate dehydrogenase.

[3] Entails shortening due to footdrop plus loss of loading during contraction of antagonists.

[4] Asterisks indicate findings consistent with that exercise advancing fast status, in that either: * the muscle became less slow than without the exercise; ** an approach toward slow status did not appear; or *** progress toward fast status was increased.

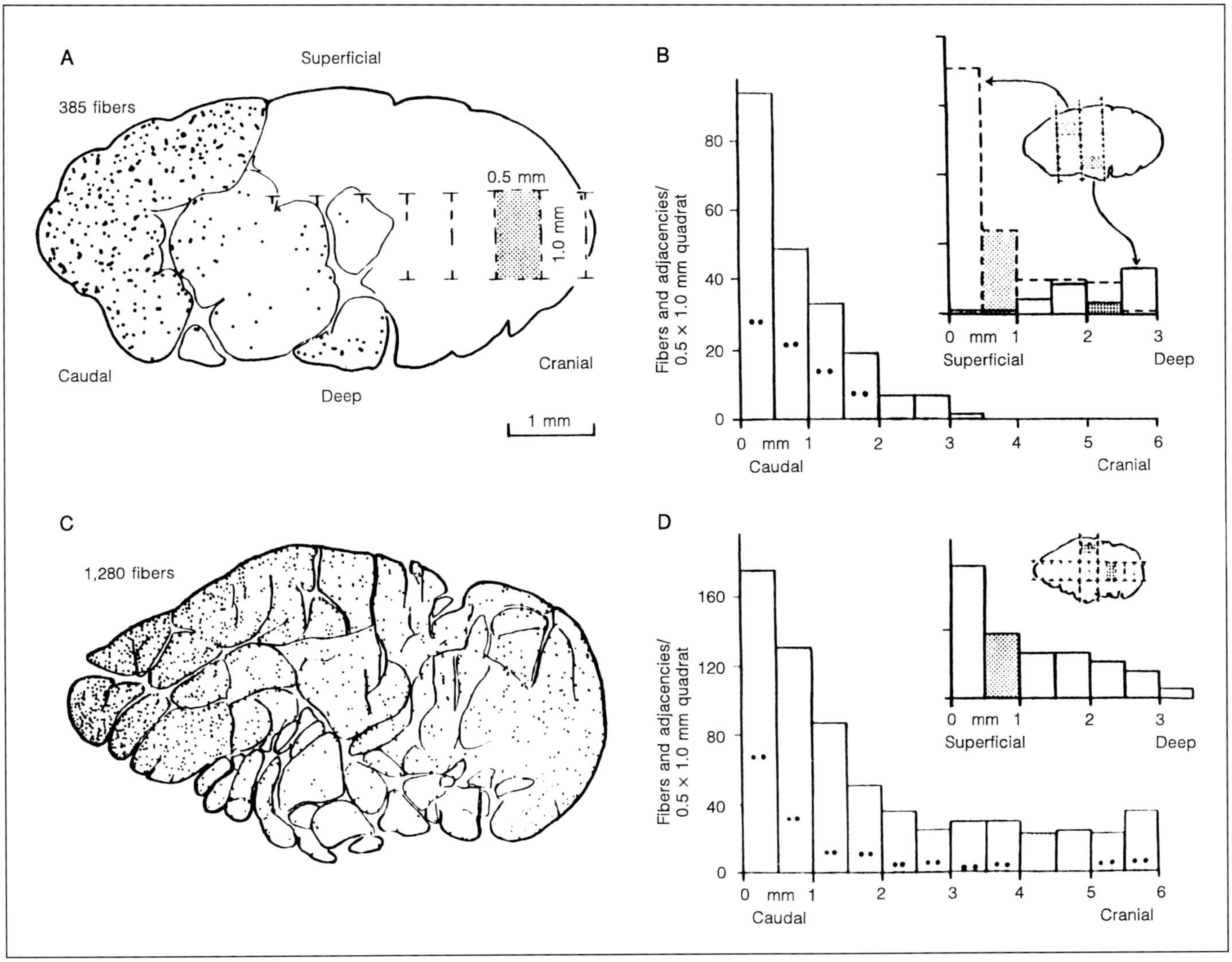

Fig. 1. Distributions of FOG fibers along the transverse and superficial-to-deep axes of cross-sections at the midlength of 2 capsularis muscles. **A** Locations of the 385 FOG fibers in one muscle, and the arrangement for counting those located along a 1.0-mm strip across the width of the muscle in 0.5 × 1.0 mm quadrats, one of which is stippled. The bars in the graph to the left (**B**) show their distribution, and paired dots the numbers of 'adjacencies' in which one FOG fiber was in contact with another. The smaller muscle outline indicates the strip down the superficial-to-deep axis at midwidth, and an additional one caudal to it along which the distributions pointed to in the adjacent bar graph were obtained. **C**, **D** Distributions in another muscle with 1,280 FOG fibers. A quadrat in each strip is stippled on the respective bar graphs and outlines.

was mounted in glycerol, and at 3 levels along its length the sarcomeres in 50-μm segments were counted. Their mean length was used to adjust that fiber's length to a sarcomere length at rest of 2.2 μm [Sacks and Roy, 1982].

Results

Fiber Types

Only SO and FOG fibers were seen in the capsularis muscles in the present and earlier [Roy et al., 1982, 1986] studies. Counts of 327, 385, 624 and 1,280 FOG fibers were obtained from 4 muscles in the present study. The mean of 654 FOG fibers is 9% that of the mean of 7,519 total fibers

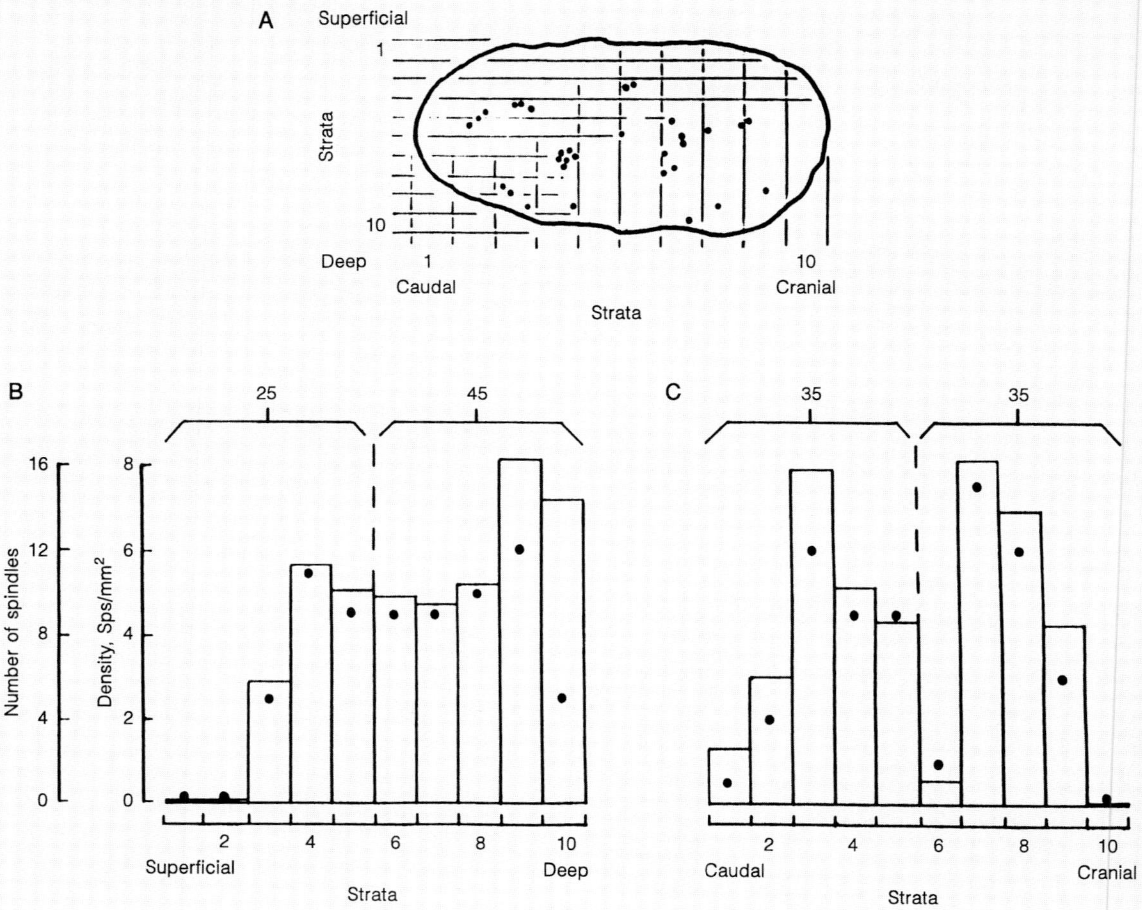

Fig. 2. Distributions of Sps through the depth and across the width of the capsularis summarized from 2 muscles. **A** Locations of all Sps in one muscle as projected onto its cross-sectional outline at midlength, and how the outline was subdivided into 10 horizontal and then 10 sagittal strata in which the Sps were counted. **B**, **C** Dots indicate the summed counts of Sps in successive strata along superficial-to-deep and transverse axes of both muscles, and the bars present the densities per area. The numbers of Sps in halves of the 2 muscles are listed above the brackets.

isolated from 2 macerated muscles [unpubl. data]. The FOG distribution along the transverse axis of the somewhat flattened outline at midlength of the muscle with 385 fibers is represented in figure 1A and B. The longer border of the trapezoidally shaped muscle is labeled 'cranial' and the other 'caudal', although in situ the muscle was directed caudo-infero-laterally. Outlined is a 1-mm strip extending across the section along which FOG fibers (dots) were counted in successive 0.5-mm wide, rectangular quadrats like the stippled one, and the bar graph to the lower left in frame B gives their incidence. On the smaller cross-sectional outline in B are indicated 2 strips extending from the 'superficial' (facing lateralward when in situ) to 'deep' (next to the ilium and joint) surfaces that were analyzed. In the graph beside the outline, those bars with solid lines give the incidence of FOG fibers in the strip centered at mid-width, and those with dashed lines the counts along the caudally adjacent strip. The distribution in the muscle with 1,280 FOG fibers is presented in frames C and D.

The bar graphs for these muscles (and the 2 others not illustrated) show an increasing cranial-to-caudal gradient in FOG content that is more definitive than the anticipated decline along the superficial-to-deep axis. Through the depth of the muscle with 385 fibers an indeterminant pattern is seen, but a clear decrease is evident in the muscle with 3 times that many FOG fibers. Toward the caudal borders of both muscles the incidence of FOG fibers and the number that contact a second FOG fiber (i.e. are in an 'adjacency') rose to levels that are high in comparison to those for territories of isolated slow motor units in the soleus and tibialis anterior [Bodine et al., 1988]. This, and the restriction of FOG fibers to one side in the muscle of frame A, suggest that MUs as well as the individual fibers were gradated in distribution.

Spindle Representation

In analyzing the distributions of Sps and to ensure recognition of each unit in tandem arrangements, a complete series of cross sections were used since some Sps did not reach the muscle midlength. The nucleated regions of 32 Sp units were detected in one muscle and 38 in the other, and the bar graph in figure 2B shows their combined distribution to increase with depth down the maximal superficial-to-deep axes of the 2 muscles. When their locations were projected to the muscle outlines at midlength, 25 were in the superficial half-areas and 45 in the deeper halves of the respective muscles. The most superficial strata lacked Sps, and then the counts rose fairly consistently until at the deepest level accentuation in curvature of the muscle outline reduced that stratum's area. Expressed as density per area of that stratum on the midlength section, its Sp content fell in line with the other strata.

The profile of Sp incidence across the capsularis, frame C, presents 2 peaks alongside a pronounced trough at midwidth. Both in the cranial and caudal longitudinal halves of the 2 muscles a combined count of 35 Sps was obtained, and the adjusted densities (as projected to the muscle midlength) were essentially the same, i.e. across the muscle's width the incidence of Sps was not correlated to SO:FOG status.

Lengths of Extrafusal Muscle Fibers and Spindles

Since it originates along a horizontal strip on the ilium and inserts on the descending femur in a cleft between heads of the vasti medius and lateralis, when extended the capsularis muscle has a flattened, trapezoidal outline. In isolating fibers in macerated muscles, the majority of fibers extended between the origin and insertion, and ended in the 'blunt' type of terminal. Consistent with this, on the surface of muscles stained en bloc for cholinesterase, motor endplates appeared in a single row near the midlength of the muscle [unpubl. data]. The lengths of representative muscle fibers isolated from 2 macerated muscles and normalized to a 2.2-μm sarcomere length, decreased fairly linearly with location across the craniocaudal axis (fig. 3). The level at which the regression line crosses the cranial border in one

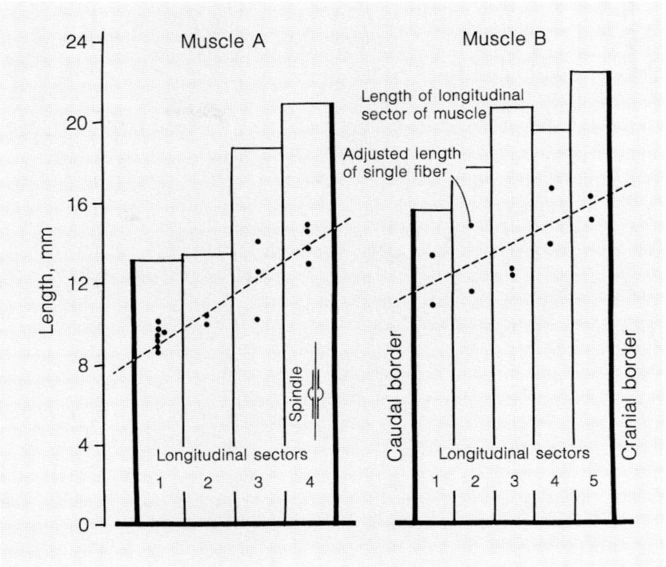

Fig. 3. Relationship of the lengths of muscle fibers isolated from 2 macerated muscles to their approximate location across the muscle width. The muscle A was subdivided into 4 equally wide longitudinal sectors and muscle B into 5, the lengths of which are indicated in the bar graphs. Lengths of representative muscle fibers isolated from each sector and normalized to a sarcomere interval of 2.2 μm are represented by dots and their relations to the order of the sectors are given by the regression lines.

muscle is 4.4 mm longer than at the caudal border, and in the other muscle 6.2 mm. At rest length, the cranial border is about 1.6 times longer than the caudal border.

In assigning lengths to Sps, a nontandem unit was measured between terminals of the axial bundle at each pole (not necessarily those of the same fiber). In a Sp linked in tandem to another, the distance from the tip of the axial bundle at the free end to the midpoint between the nucleated regions of the 2 units was measured. In a multiple linkage, the distance between 2 such midpoints was measured. In the nonextended muscle, units in the caudal and cranial halves of the muscle's breadth did not significantly differ in total Sp length, nor in the lengths of the fibrous capsule and intracapsular space (table 2). Unexpectedly, in the extended muscle Sp lengths averaged ~40% longer in the shorter, caudal than in the longer cranial half of the muscle. The means for both the capsule and intracapsular space also were greater in the caudal compared to the cranial halves, although not significantly ($p>0.05$).

Table 2. Lengths of the intrafusal fiber bundle, capsule and intracapsular space, and content of intrafusal fibers in spindles located in the cranial (longer) and caudal (shorter) half-widths of 2 capsularis muscles, one perfused with the femur flexed at the hip and the other with it extended

	Number of spindles	Lengths, mm[1,2]			IF fibers/ spindle[1]
		bundle	capsule	space	
Muscle at short length					
Cranial half	18	4.75±0.43	2.50±0.11	1.02±0.09	6.41±0.27
Caudal half	20	4.52±0.34	2.28±0.21	1.05±0.07	5.90±0.28
		n.s.	n.s.	n.s.	n.s.
Extended muscle					
Cranial half	17	5.58±0.39	3.06±0.21	1.41±0.13	5.65±0.26
Caudal half	15	7.96±0.76	3.68±0.22	1.95±0.19	6.00±0.27
		*	n.s.	*	n.s.

[1] Mean±SEM. The significance of differences between means was determined using an analysis of variance, an asterisk indicating significance at $p \leq 0.05$, and 'n.s.' as not significant. In the extended muscle the p value for capsule length was 0.1.

[2] A Sp linked in tandem to another one was considered to end at half the distance separating the centers of the 2 nucleated regions.

Discussion

Studies of the effects on contractile behavior resulting from long-term experimental procedures [Roy et al., 1986, 1992] include observations on the capsularis showing that: (1) after thoracic spinal cord transection, treadmill exercise deterred the conversion from SO to FOG fiber type; (2) after transection of the spinal cord at thoracic and sacral levels plus bilateral dorsal rhizotomy between those levels, the reductions in representation of SO fibers and their cross-sectional areas (CSAs) were 3 times greater than after spinalization alone; and (3) the ratio of fiber types after such spinal cord isolation was not influenced by manipulative exercise of the limb. The steps implicated as having detracted from slow status, i.e. the elimination of supraspinal influence on the capsularis motor pool and exclusion of all proprioceptive inflow, would have involved any Sp-mediated trophic influence on the motoneurons. Analyses of the distributions of Sps and fiber types support this possibility, although the rationale is not simple.

Superficial-to-Deep Distributions of Spindles and Fiber Types

The distribution pattern discernible in the predominantly slow capsularis and more obvious in muscles of mixed fiber type [Voss, 1936; Botterman et al., 1978; Peck et al., 1984], wherein Sps are better represented deep within a muscle where SO fibers also are more numerous, is initiated with entry of the nerve into the muscle's primordium. Each sensory axon (to become the group-I caliber fiber) follows in intimate association with a motor axon [Tosney and Landmesser, 1985] that contacts myotubes destined to become the slow extrafusal (EF) fibers that guide addition of more EF fibers [Ashmore et al., 1972; Gates and Betz, 1993], as well as those myotubes that under Ia influence form the slow bag$_2$ intrafusal (IF) fiber central to formation of the other IF fibers [Zelena, 1964; Kucera and Walro, 1994]. Neural contact with the myotubes is not requisite for the initial differentiation of EF fiber types, since this appears in the absence of a motor nerve [Condon et al., 1990], but later characterization of the emerging population of myotubes is strongly influenced by neural, to include sensory, activity. When the rat spinal cord was cut 2 h after birth [Rubinstein and Kelly, 1978] or the hindlimbs were unloaded (hindlimb suspension) at 18 days of age [Elder and McComas, 1987], the soleus did not evolve into a slow muscle, although in muscles of mixed fiber type the ratio of slow fibers was normal. In addition, in kittens spinalized at 2 weeks of age, when soleus contraction time is transiently shortening before entering a phase of progressive lengthening, the soleus remained fast at 6 months [Smith et al., 1993], and the enzymatic characterization of the muscle fibers approached that observed in fast muscles [Hoffmann et al., 1990].

For Sps to contribute to the emergence and later maintenance of the SO-fiber gradient down the superficial-to-deep axis in a muscle: (1) their incidence must increase with depth; (2) the diameter of a MU territory must be less than the distance between the muscle surfaces; (3) topographical specificity in central connections must be adequate, and (4) the motor axons must be capable of influencing fiber type. The capsularis presented a gradient in Sp representation that overall was appropriate, and the equivalent diameter of a MU was estimated as half the distance between the muscle surfaces [unpubl. observation]. That localization in reflex influence could be accommodated in so small a muscle is suggested by the restriction of Sps to a deep, central compartment of slow fibers seen in muscles of the rat and guinea pig [Yellin, 1969; Grim, 1972; Maier et al., 1976]. A relationship between the medial-lateral locations of motoneurons with the motor pool and their territories in the periphery need not stand out, judging from the failure to detect this relationship in fast and slow MUs in the cat medial gastrocnemius upon stimulation of individual motoneurons across the medial-to-lateral axis of the motor pool [Burke and Tsairis, 1973, though possibly in the dorsoventral direction motoneurons to slow motor units become more common, cf. their fig. 5].

Craniocaudal Distributions

The craniocaudal gradient of FOG fibers in which whole MUs seemed to participate implies topographic localization in innervation across the 6-mm width of the capsularis. A gradient exists among ventral (and dorsal) spinal rootlets in the medial gastrocnemius [Swett and Eldred, 1959; Swett et al., 1970] and within the motor nucleus in the lateral gastrocnemius [Weeks and English, 1985] of the cat. In addition, a correlation between the size of α-mononeuron soma along the craniocaudal extent of the motor pool and the fiber-type distribution of the medial gastrocnemius has been detected [Burke et al., 1977]. In the smaller lateral gastrocnemius of the rat, segregation of efferents from adjacent segmental levels of the cord was demonstrated by selectively sectioning ventral roots [Bennett and Lavidis, 1984], and even in the compartmentalized extensor digitorum longus segmentotopic organization was found [Balice-Gordon and Thompson, 1988]. However, in the rabbit soleus which like the capsularis is composed almost entirely of slow fibers and is not structurally compartmentalized, no correlation between the craniocaudal location of motoneurons and their territories in the muscle was detected upon intraneuronal stimulation or retrograde labeling in the neonatal period or as polyinnervation was being eliminated [Cramer and Van Essen, 1995]. Later, a spatial gradient in the ratio of fast and slow fibers across the rabbit (and rat) soleus muscles became apparent [thesis by H. Gordon quoted in Cramer and Van Essen, 1995]. In the cat's capsularis, the presence of a transverse gradient in fiber types implies that its motor supply, at least after leaving the spinal cord, is topographically arranged and available for localization in reflex and trophic influence.

The distribution found among Sps, however, did not accord with the hypothesis that proprioceptive input strengthens representation of SO MUs. Their incidence (as inferred from counting FOG fibers) rose toward the muscle's longer border, but the distribution of Sps was bimodal, with the combined totals for cranial and caudal halves of the muscles happening to be identical. Descriptions of other muscles composed largely or entirely of slow fibers, the soleus of the rat [Yellin, 1969], guinea pig [Maier et al., 1976] and cat [Arendt and Asmussen, 1974], as well as the 92%-fast [Gates and Batz, 1993] lumbrical muscles in rats [Yellin, 1969] mention that Sps appear equitably distributed across their width. In these muscles and the capsularis perhaps gradation in fiber types is not laid down during early development, but results later from influences emanating from the muscle itself, as suggested by the findings in the rabbit soleus.

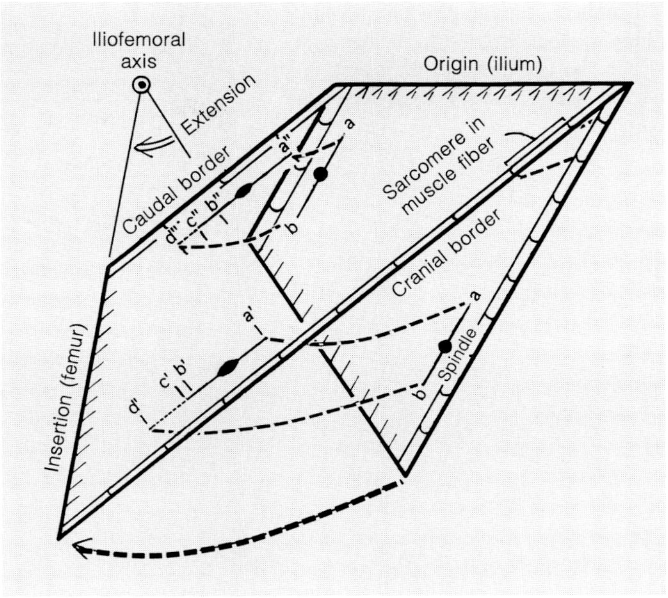

Fig. 4. Diagrammatic representation of the changes in length of muscle fibers, their sarcomeres, and of the Sps in the trapezoidally shaped capsularis as predicted to occur upon backward rotation of the femur at the right hip joint. The ratio of the muscle's dimensions are not those of the actual muscle. During elongation of the muscle, Sps near the long cranial and shorter caudal borders of the muscle, like the sarcomeres in adjacent muscle fibers, would undergo equal lengthening in proportion to their initial lengths (a–b to a'–b' = a"–b") if compliance along the Sp axis were equal to that along an equivalent length of septum and muscle that relay stress to that Sp. If compliance in the Sps were 50% *less*, differing degrees of extension would result as indicated by the intervals b'–c' and b"–c", the Sp at the *shorter* muscle border undergoing *greater* lengthening. These were calculated from the initial ratio of Sp length to that of in-series tissue, and the difference in lengthening of the muscle margin nearby upon rotation of the femur.

Table 3. Percentages of slow-oxidative fibers and their cross-sectional areas in a deep region of the capsularis muscle in control, spinal, and spinal-isolated adult, female cats[1]

	Content of SO fibers, %			Fiber cross-sectional area, μm^2		
	control	operated	ratio	control	operated	ratio
Spinal[2]	93±1	80±4	0.85	2,874±384	2,502±303	0.87
Spinal isolated[3]	96±1	50±13	0.52	2,733±341	1,481±155	0.54

[1] Values are means±SEM adapted from data on the effects of spinal cord transection [Roy and Acosta, 1986] and spinal cord isolation [Roy et al., 1992] in sample areas containing ~200 fibers. Spinalization consisted of transection of the spinal cord at T12–T13. Spinal cord isolation included transections of the cord at T12–T13 and L7–S1, plus bilateral, intradural dorsal rhizotomy between the transection levels. Observations were made ~6 months after operation.
[2] Data from 6 control and 6 operated cats.
[3] Data from 6 control and 7 operated cats.

Relation of Muscle Shape to Spindle Discharge

The bilateral balance in content of Sps need not exclude the presence of a gradated craniocaudal afferent influence. The caudal border of the capsularis crosses the iliofemoral joint ~4 mm distant to its axis of flexion and the parallel, longer border is 6 mm further forward, so that upon extending the thigh successive muscle fibers passing between the diverging lines of origin on the ilium and insertion on femur must undergo increments in length that increase linearly toward the cranial border (fig. 4; the length of a given fiber would rise as a cosine function of the joint angle). During extension of the femur, a muscle fiber at the cranial margin might lengthen 50% more than one nearer the joint axis, yet its individual sarcomeres not be subject to greater or faster excursion. Thus, during growth the addition of sarcomeres at the ends of the fibers apparently kept their lengths within the physiological range of contraction.

Spindles showed no such inbuilt compensation, for in the non-extended muscle the lengths of the IF fiber bundle, capsule and intracapsular space did not significantly differ between cranial and caudal longitudinal halves of the muscle, nor did their IF fiber content (table 2). The ratio of Sp length (e.g. the mean for nontandem Sps in the unstretched muscle) to length of the muscle border (the mean for the 2 muscles in fig. 3) would range from 1:4.8 at the longer border to 1:2.8 at the caudal one, a rise of 72%. Nevertheless, if each Sp presented the same degree of compliance as did an equivalent length of tissue in series, when the femur rotated backward the *increments* in Sp length, like those of sarcomeres in the EF fibers, would not increase with location across the muscle width (fig. 4). Each aspect of a lengthening stimulus, i.e. static length, velocity, and acceleration of change in length, should be the same across the muscle width. The gradient in fiber type across the capsularis is not attributable to an effect on Sp outflow related to its geometric configuration per se.

In the extended muscle, the length of Sps in the shorter caudal half of the muscle was significantly longer than in the cranial half (table 2). This is an indication that compliance along the axial bundle differed from that of an equivalent length of septal and endomysium tissue in series, if as the geometry indicates during rotation of the femur the stress buildup is equal across the muscle width. If elastic resistance presented by the Sp were greater, the resulting deficit in lengthening of a Sp located near the shorter border must be compensated for by additional lengthening distributed over a lesser extent of in-series tissue than in a Sp located near the long border of the muscle. Near the hip joint resistance confronting the Sp would rise higher, undergo more strain, and discharge from its endings rise faster and farther, as diagrammed in figure 4. The Sp at the short muscle border has one equivalent length of septal tissue in series, and a Sp at the cranial border has four. During backward rotation of the femur, the length (a–b) of these Sps, like that of sarcomeres depicted in nearby muscle fibers, would undergo equal lengthening from the rest position (b'–d' = b"–d"), providing that yielding along a Sp were the same as that in an equivalent length of septal and muscle tissue functionally in series. If instead, linear elastic compliance along each Sp were half that of the tissue in series (i.e. the Sp was stiffer), lengthening in both Sps would be reduced, but not equally, i.e. in the caudal Sp from b' to c', and in the cranial one only from b" to c". That Sps in the caudal half do undergo more lengthening is evident from the greater length of their axial bundles in the stretched muscle (table 3). Discharge from the Ia and II afferents must differ proportionately if their respective static sensitivities are linear.

In predicting effects on Sp discharge introduced by compliance, attention should to be directed to the 'sensory region' on the IF bundle, for it is the ratio of compliance along this segment to that of an equivalent length of tissue in series, including striated portions of the IF fibers that would distinguish responses to stretch. Along that region myofibrillae are replaced by bags of vesicular nuclei, making the fibers appear highly compliant, but cinematographic analysis has shown it to be up to 8 times stiffer than passive poles of the same bag fibers [Poppele et al., 1979], attributable to resistance of the sensory wrappings, actin-like filaments with the endings, and along the interface between them and the IF fibers (but not to elastic fibrils which congregate alongside poles of the fibers [Gladden, 1974]. The sensory region is less than a tenth as long as the entire axial bundle, but in Sps located along cranial and caudal borders of the muscle the ratio of length of the nucleated region to that of tissue in series would differ by the same multiplication factor, and the interpretation made from observations on the entire axial bundle still applies.

The muscle's trapezoidal shape could lead to other differentials in stress. Over the distal surface an aponeurosis extends upward about the same distance at all levels across the muscle width, so the proportion of muscle overlaid increases toward the shorter border. Stretch responses of Sps located beneath an aponeurosis are damped [Meyer-Lohmann et al., 1974], implying that beyond it Sps lengthen disproportionately, and most Sps in the capsularis are so located. Also related to resistance in parallel is a tendency for endo- and perimysium to contain more collagen fibers in a slow muscle or region within a muscle where slow MUs predominate [Kovanen et al., 1984]. When activated the muscle fibers themselves are less compliant in slow than fast MUs [Petit et al., 1990], and subsequent to activation present more short-range stiffness [Malamud et al., 1996; Mutungi et al., 1996]. Thus, at the cranial border, where the incidence of slow fibers is greater, Sps should be subject to less stress. Any gradient in response, however produced, would be exaggerated during fusimotor activation. Position sensitivities both of Ia and II afferents are greater when the Sp's IF fibers are stiffened [Botterman and Eldred, 1982; Poppele and Quick, 1985].

Mediation of Fiber-Type Transition along the Craniocaudal Axis

If the discharge from Sps contributes to SO status and the volume of discharge is higher toward the hip joint, a higher incidence of SO fibers should be found in this region. Actually FOG fibers were at their maximum. In the muscle in figure 1A, the cranial half had no FOG fibers, implying that the level of Sp discharge from that region was optimal for support of SO fibers. Perhaps near the short border the discharge attained during stretch exceeded that level, and promoted conversion of MUs to FOG status. Chronic stimulation of a muscle nerve at rates of 5–20 Hz results in fast fibers converting towards a slow status [Salmons and Henrikssen, 1981; Brown et al., 1989; Jarvis et al., 1996], and after spinalization [Harada, 1983], spinalization plus tenotomy [Salmons and Vrbova, 1969], cross-innervation with a fast muscle nerve [Harada, 1983], or hindlimb suspension [Leterme and Falempe, 1994; Canon, 1995] retarded the transition to a faster status. Direct stimulation of the rabbit soleus at 5–10 Hz, but not at 20–40 Hz, following spinalization plus tenotomy prevented it from becoming fast [Salmons and Vrbova, 1969]. Similarly, stimulation of the denervated rat soleus at 10 or 15 Hz led to continuance of its slow contractions, but at 100 Hz reduced the contraction time [Gorza et al., 1988]. Stimulation at 150 Hz maintained normal characteristics of the denervated extensor digitorum brevis [Eken and Gundersen, 1988], and made the soleus faster [Gundersen et al., 1988]. Even in an intact animal, stimulation at high frequencies can convert slow fibers to a faster status [Lomo et al., 1974; Pette and Vrbova, 1985]. Excessive loading of antigravity muscles by long-term centrifugation has a similar effect [Roy et al., 1996].

Parameters of Inflow from Spindles

Although distinction of fiber types can appear in the fetus [Condon et al., 1990] or later be induced without neural mediation, when the motor axon is intact it does provide for efficient dissemination of the digital discharge pattern and trophic influence characteristic of that type of motoneuron. Thus, after a slow muscle was cross-innervated with the nerve to a fast muscle, its contractions became faster [Buller et al., 1960b], troponins of fast and slow muscles approached a common level [Eldridge et al., 1984], and slow tonic muscle in the newly hatched chick showed a shift in myosin type [Syrovy and Zelena, 1975]. Even minced fast or slow muscle tissue took on characteristics promoted by the nerve that normally supplies the host site [Gordon et al., 1977; Snoj-Cvetko et al., 1996]. Reflex afferent influence is not essential, since after cross-innervation of the soleus with the nerve from a fast muscle, progression toward fast status was not detectably different if the limb was deafferented [Luff et al., 1984]. Nevertheless, motoneurons are malleable, as illustrated by the reduced loss in after-hyperpolarization of soleus motoneurons that follows spinalization [Cope et al., 1986], and fast-to-slow conversion of innate motoneuron properties following low-frequency activation [Munson et al., 1997]. Retrograde influence subject to length and mobility status of the muscle seems to be involved, for the reduced loss still occurs if dorsal roots to the soleus are also sectioned if the muscle was kept at a shortened length by casting the limb with the ankle dorsiflexed, but did not if it was lengthened or not immobilized [Gallego et al., 1979]. In an intact animal, the length and activity status of the soleus, of course, is subject to a strong reflex influence from Sp inflow.

Long-term procedures that reduce or enhance proprioceptive input, or differentially affect the steady or dynamic nature of Sp discharge are often associated with contrasting shifts in contractile status (table 1), suggesting that Sp input does inflect trophic influence from motoneurons. This is consistent with the increases in incidence of slow fibers and Sps with increasing depth through the capsularis. Across the muscle width the number of FOG fibers was unrelated to the sheer incidence of Sps, but a gradient in influence might emerge if responses to stretch varied with location of Sps. As the discussion above on compliance emphasized, the parameter of Sp discharge subject to gradation might be quantitative: (1) the instantaneous firing rate of the individual Sp [Buller et al., 1960a], or (2) more relevant to the population of Sps' an 'amount effect' [Gundersen et al., 1988; Lomo et al., 1989], wherein contractile status is related to the accumulation of stimulus pulses. Alternatively, the key parameter might be qualitative: the pattern of axonal discharges, i.e. timing of spike trains, pairing and cresting of initial discharges [Eken and Gundersen, 1988] that distinguish the continuous neural activity associated with slow muscles from the episodic discharges to fast muscle. Related to all three parameters is the balance of inflow from primary and secondary Sp afferents.

The rate of discharge of a passive group II Sp afferent rises more rapidly with increase in static length of the muscle than does that from a Ia ending [Matthews, 1972; Botterman and Eldred, 1982; Poppele and Quick, 1985], and emphasis in connections directly to motoneurons and indirectly through interneurons differs [Sypert et al., 1980]. Fiber status might be influenced by the ratio of the Ia and II inputs, those procedures giving rise to continuous, low-frequency discharge (in table 1, 'static loading') fostering slow contractility, and fast status emerging if such an influence is shifted by 'unloading', or at the extreme, spinal cord isolation. If the muscle participates in more dynamic events, Ia behavior and connections would dominate and might strengthen fast motoneuronal status, as suggested by some procedures that increase phasic Sp input (asterisks under 'dynamic exercise' in table 1). For instance, during long-term suspension of the hindlimbs a failure to prevent conversion of soleus fibers to FOG status by daily exercise consisting of dropping rats onto their hindlimbs [Hauschka et al., 1987] may have been due to conversion to faster MU status by the extremes in Ia discharge aroused during abrupt, eccentric contractions. Conversely, the reduction in conversion to fast status during treadmill exercise of spinalized cats [Roy and Acosta, 1986] could have resulted because with the moderate excursions and more enduring antigravity activation the balance in afferent type input favored slow status. In intact cats kept under barbiturate anesthesia, the suppression of movements and fusimotor activity that particularly bolstered Ia input may have contributed to slowing of the fast flexor digitorum longus [Davis and Montgomery, 1977].

Together the topographical relations found of Sps to fiber types, and measurements indicating the distribution of stress on Sps in the capsularis support the hypothesis that afferent input from Sps affects fiber-type status if 2 modes of influence are recognized: (1) down the superficial-to-deep axis, the gradation in incidence of Sps, and (2) across the width of this trapezoidal muscle, a mechanical differential affecting the volume or some other parameter of proprioceptive discharge. Despite the muscle's size the necessary topographical localization in afferent and motor flows apparently is available. In muscles having more fiber types, compartmentalization, serial fiber arrangement, etc., this relationship of Sps and muscle fibers would be less evident, but Sp input would still have an influence, at least on the related SO and FOG types of fibers.

Acknowledgements

The contributions of Sharon Sampogna, Staff Research Associate, in preparing the histological material, and Daniel Eldred for reviewing the geometrical arguments are deeply appreciated.

This work was supported in part NIH grant NS 16333.

References

Ansved, T. (1995) Effects of immobilization on the rat soleus muscle in relation to age. Acta Physiol Scand *154:* 291–302.

Arendt, K.-W., G. Asmussen (1974) Enzymhistochemische Untersuchungen an Muskelspindeln verschiedener Species. Anz Bd *136:* 213–228.

Ashmore, C.R., D.W. Robinson, P. Rattray, L. Doerr (1972) Biphasic development of muscle fibers in the fetal lamb. Exp Biol *37:* 241–255.

Baldwin, K.M., R.R. Roy, R.D. Sacks, C. Blanco, V.R. Edgerton (1984) Relative independence of metabolic enzymes and neuromuscular activity. J Appl Physiol *56:* 1602–1607.

Baldwin, K.M., V. Valdez, R.E. Herrick, A.M. MacIntosh, R.R. Roy (1982) Biochemical properties of over-loaded fast-twitch skeletal muscle. J Appl Physiol *52:* 467–472.

Balice-Gordon, R.J., W.J. Thompson (1988) The organization and development of compartmentalized innervation in rat digitorum longus muscle. J Physiol (Lond) *398:* 211–231.

Barnard, R.J., V.R. Edgerton, J.B. Peter (1970) Effect of exercise on skeletal muscle. I. Biochemical and histochemical properties. J Appl Physiol *28:* 762–766.

Bennett, M.R., N.A. Lavidis (1984) Development of the topographical projection of motor neurons to a rat muscle accompanies loss of polyneuronal innervation. J Neurosci *4:* 2204–2212.

Bodine, S.C., A. Garfinkel, R.R. Roy, V.R. Edgerton (1988) Spatial distribution of motor unit fibers in the cat soleus and tibialis anterior muscles. J Neurosci *8:* 2142–2152.

Botterman, B.R., M.D. Binder, D.G. Stuart (1978) Functional anatomy of the association between motor units and muscle receptors. Am Zool *18:* 135–152.

Botterman, B.R., E. Eldred (1982) Static stretch sensitivity of Ia and II afferents in the cat gastrocnemius. Pflügers Arch *395:* 204–211.

Boyes, G., I. Johnston (1979) Muscle fiber composition of rat vastus intermedius following immobilization at different muscle lengths. Pflügers Arch *381:* 195–200.

Brown, J.M.C., J. Henrikkson, S. Salmons (1989) Restoration of fast muscle characteristics following cessation of chronic stimulation: Physiological, histochemical and metabolic changes during slow-to-fast transformation. Proc R Soc Lond [B] *235:* 321–346.

Buller, A.J., J.C. Eccles, R.M. Eccles (1960a) Differentiation of fast and slow muscles in the cat hind limb. J Physiol (Lond) *150:* 399–416.

Buller, A.J., J.C. Eccles, R.M. Eccles (1960b) Interactions between motoneurones and muscles in respect to the characteristic speeds of their responses. J Physiol (Lond) *150:* 417–439.

Burke, R.E., P.L. Strick, K. Kanda, C.C. Kim, B. Walmsley (1977) Anatomy of medial gastrocnemius and soleus motor nuclei in cat spinal cord. J Neurophysiol *40:* 667–680.

Burke, R.E., P. Tsairis (1973) Anatomy and innervation ratios in motor units of cat gastrocnemius. J Physiol (Lond) *234:* 749–765.

Caiozzo, V.J., M.J. Baker, R.E. Herrick, M. Tao, K. Baldwin (1996) Effect of spaceflight on skeletal muscle. J Appl Physiol *76:* 1764–1773.

Canon, F., A.X. Bigard, D. Merino, F. Lienhard, C.Y. Guezennec (1995) Effects of chronic stimulation on structural and metabolic properties of hindlimb suspended rat soleus muscle. Eur J Appl Physiol Occup Physiol *70:* 528–535.

Condon, K., L. Silberstein, L. Blau, H.M. Thompson (1990) Differentiation of fiber types in aneural musculature of the prenatal rat forelimb. Dev Biol *138:* 275–295.

Cope, T.C., S.C. Bodine, M. Fournier, V.R. Edgerton (1986) Soleus motor units in chronic spinal transected cats: Physiological and morphological alterations. J Neurophysiol *55:* 1202–1220.

Cramer, K.S., D.C. Van Essen (1995) Lack of topography in the spinal cord projection to the rabbit soleus muscle. J Comp Neurol *35:* 404–414.

Davis, C.J.F., A. Montgomery (1977) The effect of prolonged activity upon the contraction characteristics of fast and slow mammalian twitch muscle. J Physiol (Lond) *270:* 581–594.

Edgerton, V.R., L. Gerchman, R. Carrow (1969) Histochemical changes in rat skeletal muscle after exercise. Exp Neurol *24:* 110–123.

Eken, T., K. Gundersen (1988) Electrical stimulation resembling motor unit activity: Effects on denervated fast and slow rat muscles. J Physiol (Lond) *402:* 651–669.

Elder, G.C.B., A.J. McComas (1987) Development of rat muscle during short- and long-term hindlimb suspension. J Appl Physiol *62:* 1917–1923.

Eldridge, L., G.K. Dhoot, W.F.H.M. Mommaerts (1984) Neural influences on the distribution of troponin I isotopes in the cat. Exp Neurol *83:* 328–348.

Engel, W.K., M. Brooke, P.G. Nelson (1966) Histochemical studies of denervated or tenotomized cat muscle: illustrating difficulties in relating experimental animal conditions to human neuromuscular diseases. Ann NY Acad Sci *138:* 160–185.

Fischbach, G.D., N. Robbins (1969) Changes in contractile properties of disused soleus muscles. J Physiol (Lond) *201:* 305–320.

Gallego, R., M. Kuno, R. Nunez, W.D. Snider (1979) Dependence of motoneurone properties on the length of immobilized muscle. J Physiol (Lond) *291:* 179–189.

Gates, H.J., W.J. Betz (1993) Spatial distribution of muscle fibers in a lumbrical muscle of the rat. Anat Rec *236:* 381–389.

Gladden, M.H. (1974) Elastic fibres in muscles spindles of the cat. J Physiol (Lond) *227:* 45–46P.

Gonyea, H., D.C. Van Essen (1985) Specific innervation of muscle fiber types in a developmentally polyinnervated muscle. Dev Biol *111:* 42–50.

Gordon, T., R. Perry, T. Srihari, G. Vrbova (1977) Differentiation of slow and fast muscles in chickens. Cell Tiss Res *180:* 211–222.

Gorza, L., K. Gundersen, T. Lomo, S. Schiaffino, R.H. Westgaard (1988) Slow-to-fast transformation of denervated soleus muscles by chronic high-frequency stimulation in the rat. J Physiol (Lond) *402:* 627–649.

Graham, S.C., R.R. Roy, E.O. Hauschka, V.R. Edgerton (1989a) Effects of periodic weight support on medial gastrocnemius fibers of suspended rats. J Appl Physiol *67:* 945–953.

Graham, S.C., R.R. Roy, S.P. West, D. Thomason, K. Baldwin (1989b) Exercise effects on the size and metabolic properties of soleus fibers in hindlimb-suspended rats. Aviat Space Med *60:* 226–234.

Grim, M. (1972) On the problem of irregular distribution of muscle spindles. Folio Morphol *20:* 119–120.

Gundersen, K., E. Leberer, T. Lomo, D. Pette, R.S. Staron (1988) Fibre types, calcium-sequestering, and metabolic enzymes in denervated and chronically stimulated muscles of the rat. J Physiol (Lond) *398:* 177–189.

Guth, L., J.B. Wells (1972) Physiological and chemical properties of the soleus muscle after denervation of its antagonists. Exp Neurol *36:* 463–471.

Harada, Y. (1983) Experimental study of denervated rat muscle. II. The effects of electrical stimulation on the denervated rat muscles. J Jpn Orthop Assoc *57:* 859–866.

Hauschka, E.O., R.R. Roy, V.R. Edgerton (1987) Size and metabolic properties of single muscle fibers in rat soleus after hindlimb suspension. J Appl Physiol *62:* 2338–2347.

Herbert, M.E., R.R. Roy, V.R. Edgerton (1988) Influence of one-week hindlimb suspension and intermittent high-load exercise on rat muscles. Exp Neurol *102:* 190–198.

Hoffmann, S.J., R.R. Roy, C.R. Blanco, V.R. Edgerton (1990) Enzyme profiles of single muscle fibers never exposed to normal neuromuscular activity. J Appl Physiol *69:* 1150–1158.

Jarvis, C.J., T. Mokrusch, M.M.N. Kwende, H. Sutherland, S. Salmons (1996) Fast-to-slow transformation in stimulated rat muscle. Muscle Nerve *19:* 1469–1480.

Jiang, B., R.R. Roy, V.R. Edgerton (1990) Enzymatic plasticity of medial gastrocnemius fibers in the adult chronic spinal cat. Am J Physiol *259:* C507–C514.

Johnson, D.J., L.A. Smith, E. Eldred, V.R. Edgerton (1982) Exercise-induced changes of biochemical, histochemical, and contractile properties of muscle in cordotomized kittens. Exp Neurol *76:* 414–427.

Kovanen, V., H. Suominen, E. Heikinen (1984) Collagen of slow twitch and fast twitch muscle fibers in different types of rat skeletal muscle. Eur J Appl Physiol *52:* 235–242.

Kucera, J., J.M. Walro (1994) Sequences of intrafusal fiber formation are muscle dependent in rat hindlimbs. Anat Embryo *190:* 273–286.

Leterme, D., M. Falempin (1994) Compensatory effects of chronic electrostimulation on unweighted rat soleus muscle. Pflügers Arch *426:* 155–160.
Lomo, T.R. (1989) Long-term effects of altered activity on skeletal muscle. Biomed Biochim Acta *48:* S432–S444.
Lomo, T., R.H. Westgaard, H.A. Dahl (1974) Contractile properties of muscle: control by pattern of muscle activity in the rat. Proc Roy Soc Lond [B] *187:* 99–103.
Luff, A.R., U. Proske, S.N. Webb (1984) The transformation of cross-innervated slow-twitch muscle after deafferentation in the cat. Exp Brain Res *55:* 152–157.
Maier, A., E. Eldred, V.R. Edgerton (1972) The effects on spindles of muscle atrophy and hypertrophy. Exp Neurol *37:* 100–123.
Maier, A., D.R. Simpson, V.R. Edgerton (1976) Histological and histochemical comparisons of muscle spindles in three hind limb muscles of the guinea pig. J Morphol *148:* 185–192.
Malamud, J.G., R.E. Godt, T.R. Nichols (1996) Relationship between short-range stiffness and yielding in type-identified, chemically skinned muscle fibers from the cat triceps-surae muscles. J Neurophysiol *76:* 2280–2289.
Matthews, P.B.C. (1972) Mammalian Muscle Receptors and their Central Actions. Baltimore, Williams & Wilkins, pp 140–194.
Meyer-Lohmann, J., W. Riebold, D. Rorecht (1974) Mechanical influence of the intrafusal muscle on the static behaviour of deefferented primary muscle spindle endings in the cat. Pflügers Arch *352:* 267–278.
Munson, J.B., R.C. Foehring, L.M. Mendell, T. Gordon (1997) Fast-to-slow conversion following chronic low-frequency activation of medial gastrocnemius muscle in cats. II. Motoneuron properties. J Neurophysiol *77:* 2605–2615.
Mutungi, G., K.W. Ranatunga (1996) The viscous, viscoelastic and elastic characteristics of resting fast and slow mammalian (rat) muscle fibres. J Physiol (Lond) *496:* 827–835.
Peck, D., D.F. Buxton, A. Nitz (1984) A comparison of spindle concentrations in large and small muscles acting in parallel combinations. J Morphol *180:* 243–252.
Peter, J.B., R.J. Barnard, V.R. Edgerton, C.A. Gillespie, K.E. Stempel (1972) Metabolic profiles of three fiber types of skeletal muscle in guinea pigs and rabbits. Biochemistry *11:* 2627–2633.
Petit, J., G.M. Filippi, F. Emonet-Denand, C.C. Hunt, Y. Laporte (1990) Changes in muscle stiffness produced by motor units of different types in peroneus longus muscle of cat. J Neurophysiol *63:* 190–197.
Pette, D., G. Vrbova (1985) Neural control of phenotype expression in mammalian muscle fibers. Muscle Nerve *8:* 676–689.
Poppele, R.E., D.C. Quick (1985) Effect of intrafusal muscle mechanics on mammalian muscle spindle sensitivity. J Neurosci *5:* 1881–1885.
Poppele, R.E., W.R. Kennedy, D.C. Quick (1979) A determination of static mechanical properties of intrafusal muscle in isolated cat muscle spindles. Neuroscience *4:* 401–411.
Roy, R.R., L. Acosta, Jr. (1986) Fiber type and fiber size changes in selected thigh muscles six months after low-thoracic spinal cord transection in adult cats: Exercise effects. Exp Neurol *92:* 675–686.
Roy, R.R., I.D. Meadows, K.M. Baldwin, V.R. Edgerton (1982) Functional significance of compensatory overloaded fast muscle. J Appl Physiol *52:* 473–478.
Roy, R.R., D.J. Pierotti, V. Flores, W. Rudolph, V.R. Edgerton (1992) Fiber size and type adaptations to spinal isolation and cyclical passive stretch in cat hindlimb. J Anat *180:* 491–499.
Roy, R.R., M.E. Roy, R.J. Talmadge, R. Mendoza, R.E. Grindeland, M. Vasques (1996) Size and myosin heavy chain profiles of rat hindlimb extensor muscle fibers after 2 weeks at 2 G. Aviat Space Environ Med *67:* 854–858.
Rubinstein, N.A., A.M. Kelly (1978) Myogenic and neurogenic contributions to the development of fast and slow-twitch muscles in rat. Dev Biol *62:* 473–485.
Sacks, R.D., R.R. Roy (1982) Architecture of hindlimb muscles of cats: Functional significance. J Morphol *173:* 185–195.
Salmons, S., J. Henriksson (1981) The adaptive response of skeletal muscle to increased use. Muscle Nerve *4:* 94–105.
Salmons, S., G. Vrbova (1969) The influence of activity on some contractile characteristics of mammalian fast and slow muscles. J Physiol (Lond) *201:* 535–549.
Smith, L.A., E. Eldred, V.R. Edgerton (1993) Effects of age at cordotomy and subsequent exercise on contraction times of motor units in the cat. J Appl Physiol *75:* 2683–2688.
Snoj-Cvetko, E., J. Sketelj, I. Dolenc, S. Obreza, C. Jamot, A. d'Albis, I. Erzen (1996) Regenerated rat fast muscle transplanted to the slow muscle bed and innervated by the slow nerve exhibits an identical myosin heavy chain repertoire to that of the slow muscle. Histochem Cell Biol *106:* 473–480.
Spector, S.A. (1985) Trophic effects on the contractile and histochemical properties of rat soleus muscle. J Neurosci *5:* 2189–2196.
Swett, J.E., E. Eldred (1959) Relation between spinal level and peripheral location of afferents in calf muscles of the cat. Am J Physiol *196:* 819–823.
Swett, J.E., E. Eldred, J.A. Buchwald (1970) Somatotopic cord-to-muscle relations in efferent innervation of cat gastrocnemius. Am J Physiol *219:* 762–766.
Sypert, G.W., J.W. Fleshman, J.B. Munson (1980) Comparison of monosynaptic actions of medial gastrocnemius group Ia and II muscle spindle afferents on triceps surae motoneurons. J Neurophysiol *44:* 726–738.
Syrovy, I., J. Zelena (1975) The onset and progress of transformation of avian slow into fast muscles under neural influence. Pflügers Arch *360:* 121–134.
Talmadge, R.J., R.R. Roy, V.R. Edgerton (1995) Prominence of myosin heavy chain hybrid fibers in soleus muscle of spinal cord-transected rats. J Appl Physiol *78:* 1256–1265.
Templeton, G.H., M. Padalino, J. Manton, M. Glasberg, C.J. Silver, P. Silver, G. DeMartino, T. Leconey, G. Klug, H. Hagler, J.L. Sutko (1984) Influence of suspension hypokinesia on rat soleus muscle. J Appl Physiol *56:* 278–286.
Tosney, K.W., L.T. Landmesser (1985) Growth cone morphology and trajectory in the lumbosacral region of the chick embryo. J Neurosci *5:* 2345–2358.
Tyc, F., G. Vrbova (1995) The effect of partial denervation of developing rat fast muscles on their motor unit properties. J Physiol (Lond) *482:* 651–660.
Voss, H. (1936) Ein besonders reichliches Vorkommen von Muskelspindeln in der tiefen Portion der M. masseter des Menschen und der Anthropoiden. Anat Anz *81:* 290–292.
Weeks, O.I., A.W. English (1985) Compartmentalization of the cat lateral gastrocnemius motor nucleus. J Comp Neurol *235:* 255–257.
Yellin, H. (1969) A histochemical study of muscle spindles and their relationship to extrafusal fiber types in the rat. Am J Anat *125:* 31–46.
Zelena, J. (1964) Development, degeneration and regeneration of receptor organs. Prog Brain Res *13:* 175–213.

Acta Anatomica

Acta Anat 1997;159:127–135

J.W. Hermanson

Department of Anatomy,
College of Veterinary Medicine,
Cornell University, Ithaca, N.Y., USA

Architecture and the Division of Labor in the Extensor carpi radialis Muscle of Horses

Key Words
Muscle architecture
Histochemistry
Myosin isoforms
Horse

Abstract
The extensor carpi radialis muscle of the horse is deceptive at first appearance. It has a fusiform shape similar to other forearm extensor muscles. The fiber arrangement also appears long and relatively parallel. However, it may contain two or more compartments that correlate with differing functional roles. Histochemical and immunocytochemical analysis of proximal and distal regions of the muscle (n=9) demonstrate that the proximal portion of the muscle is composed of a mean of 13% type I, presumed slow twitch, and 61% type IIb, presumed fast twitch fibers. In contrast, the distal compartment is composed of a mean of about 43% type I and only 22% type IIb fibers. The type I and IIa fibers are all highly aerobic based on nicotinamide dinucleotide tetrazolium reductase reactions. Correlative data regarding the myosin isoforms has been obtained with 4% SDS-PAGE analysis of myosin heavy chain isoforms which demonstrate isoforms migrating at rates similar to rat type I, IIa, and IIx. The latter has been referred to as type IIB/X in a study of the horse's gluteus medius muscle. We propose that the in-series 'compartmentalization' of the muscle, while not conforming strictly to the definitions of neuromuscular compartments, relates to the insertion of the lacertus fibrosus, a distal slip of the biceps brachii, upon the extensor carpi radialis. Earlier studies demonstrated a high proportion of type I fibers in the equine lateral biceps brachii which were thought to stabilize the shoulder during long periods of quiet standing. Because of action imposed on the distal compartment by the biceps brachii, slow and fatigue-resistant functions are part of the limb's passive stay apparatus to effect long-term standing by the horse. Thus, the fatigue-resistant compartments of biceps brachii and extensor carpi radialis may constitute an in-series arrangement of the two muscles. The proximal compartment is suited to provide powerful, more fatigable contractions during locomotion and likely affects stress or strain within the distal postural compartment.

KARGER

Fax+41 61 306 12 34
E-Mail karger@karger.ch
www.karger.com

0001–5180/97/1593–0127$15.00/0

This article is also accessible online at:
http://BioMedNet.com/karger

John W. Hermanson
Department of Anatomy, College of Veterinary Medicine, Cornell University
Ithaca, NY 14853-6401 (USA)
Tel. +1 607 253 3542
E-Mail jwh6@cornell.edu

Table 1. Fiber type composition (%) for proximal and distal portions of the ECR muscles in horses

Breed	Sex	Age years	Weight kg	Proximal			Distal		
				type I	type IIa	type IIb	type I	type IIa	type IIb
Standardbred	F	20	500	9.0	35.6	55.3	45.1	54.9	0
Standardbred	F	7	485	16.2	24.7	59.1	34.8	39.9	25.3
Standardbred	F	6	400	9.3	31.3	59.4	42.2	33.0	24.8
Thoroughbred	F	10	591	7.0	18.6	74.4	38.0	21.0	41.0
Thoroughbred	F	9	490	4.9	12.9	82.2	51.3	28.1	20.6
Thoroughbred	F	8	511	31.2	31.6	37.1	49.2	32.4	18.4
Thoroughbred	F	6	545	7.9	21.9	70.0	45.9	26.5	27.6
Mean				12.2[a]	25.2[a, b]	62.5[e]	43.8[c, d]	33.7[b, c, d]	22.5[a, b]
SD				9.1	8.1	14.7	5.9	10.8	12.3

Means with different superscript letters are significantly different ($p<0.05$).

The muscles of horses have been studied with an interest in their potential function in the galloping animal. Consistent with this, studies have documented the form and function of horse muscle in terms of $\dot{V}O_2max$ [Armstrong et al., 1992]. While these studies have proven valuable in estimating the design constraints upon horses, our group has often considered the observation that many horses stand, relatively quietly, for a considerable portion of their day. As such, significant components of the forelimb musculature consist of postural muscle fibers or type I, presumed slow-twitch and fatigue-resistant fibers. For example, the lateral portion of the biceps brachii muscle is composed primarily of type I muscle fibers while the adjacent, in parallel medial portion of this muscle contains a higher proportion of presumed fast-twitch, type IIa fibers [Hermanson and Hurley, 1990; Hermanson et al., 1991]. We found this to be striking in an animal that was 'born to run', and in contrast to many published studies on the hindlimb hip extensor, the gluteus medius [see Bruce and Turek, 1985; Rivero et al., 1993; Serrano et al., 1996]. The triceps muscle mass also consisted of fast-twitch type II fibers. Postural activity, if reflected by a high proportion of type I fibers, was constrained to the small medial head of the triceps brachii and its parallel synergist, the anconeus: both were nearly 100% type I fibers [Ryan et al., 1992].

The present study was undertaken to determine the muscle fiber type composition and architecture of the large forearm extensor carpi radialis (ECR) muscle. This muscle represents a significant portion of the total muscle mass of the forearm extensor musculature and on general appearance seems to be a simple parallel fibered muscle of the classical literature. The ECR originates along the lateral epicondyle of the humerus, has a single head, and inserts upon the proximal part of the 'cannon' or third metacarpal bone [Dyce et al., 1996]. The muscle is active prior to and during early portions of the stance phase in the walking pony [Jansen et al., 1992] and thus contributes to stabilization of the horse's carpus or the 'knee' in horsemen's terms. A priori, a relatively homogeneous muscle, perhaps containing many type IIa fibers, might be predicted based on two observations: the muscle must support the carpus during high speed running, and this muscle provides the majority of muscular action effected by insertion on the dorsal aspect of the carpus and metacarpal bone. A previous study reported that the horse's extensor carpi radialis was composed of 9, 37, and 56% type I, IIa and IIb fibers, respectively [Andrews and Spurgeon, 1986]. It was of interest, therefore, to discover an impressive proximal to distal gradient in the muscle fiber type composition, including a large proportion of type I. This study demonstrated that the unique in-series relationship with the biceps brachii, and the muscle's fiber architectural patterns, was most consistent with an intramuscular division of labor facilitating both postural and dynamic functions for this muscle. Preliminary data from this study were presented by Hermanson and Cobb [1991] at the American Association of Veterinary Anatomists Conference.

Methods

Whole muscles were dissected free within 1 h of death from seven adult horses, *Equus caballus* (table 1). All horses died of natural causes unrelated to gait or musculoskeletal conditions. The most common cause of death was acute onset colic. Animals were euthanized by intravenous injection of a lethal dose of sodium pentobarbital. The muscles were blotted dry and weighed. Single samples (about 2 cm^3) of tissue were isolated from the superficial surface of the muscle at the proximal and distal thirds of the muscle. All samples were mounted on cork with 5% gum tragacanth and were quick-frozen in isopentane cooled to about –160 °C in liquid nitrogen. Samples were stored at –80 °C.

Serial transverse sections (8- to 10-µm-thick) were cut using a cryostat (International Equipment, Needham, Mass.) and mounted on glass slides for staining. Sections of each muscle were stained for myofibrillar ATPase following acidic (pH 4.3, 4.4, or 4.5 for 5 min at 21 °C) or basic (pH 10.3 for 10 min at 37 °C) preincubation and a 30-min incubation at 37 °C in a sodium barbital buffer containing ATP at a working pH of 9.4. Additional details about the buffers used are available in Hermanson and Hurley [1990]. Serial sections were stained for nicotinamide dinucleotide tetrazolium reductase (NADH-TR) after Novikoff et al. [1961] to demonstrate oxidative potential, and for α-glycerophosphate dehydrogenase (α-GPD) modified after Wattenberg and Leong [1969] to demonstrate glycolytic potential. Other serial sections were processed for immunocytochemistry to demonstrate antifast myosin (MY32, Sigma, diluted at 1:400 in 0.1 *M* PBS) and antislow myosin (S58, courtesy of Dr. Frank Stockdale, Stanford Medical School; and NCL-MHCs, Novacastra Laboratories, Newcastle upon Tyne, UK, diluted at 1:40 in PBS). Briefly, tissues for immunostaining were placed on subbed slides and rehydrated in a 2% normal goat serum in 0.5 *M* PBS. Tissues were incubated for 16 h in the primary antibodies (control slides were incubated in 2% normal goat serum) in a humid chamber at 4 °C. Tissues were then rinsed 3 times in PBS and incubated for 10 min in a goat antimouse biotinylated secondary antibody and processed using a Histostain SP kit (Zymed Laboratories, San Francisco, Calif.) using an AEC chromagen staining technique.

Histochemical and morphometric analysis was carried out as described in Hermanson and Cobb [1992]. Differences in fiber type proportions or size were assessed using ANOVA and a post hoc test (Tukey's) for comparison of multiple means. Data from a 20-year-old horse were excluded from these comparison tests because of the absence of type IIb fibers in the distal compartment, and because the age of the animal made it an outlier. Significance was ascribed if $p<0.05$.

Dissections were performed on two fresh ECR muscles to determine intramuscular nerve distributions. Whole muscles were also obtained from two adult horses for treatment in 40% nitric acid and analysis of muscle fiber architecture following Hermanson and Cobb [1992]. This facilitated measurement of fiber bundle lengths in various regions of the muscle.

Results

Fiber Type Composition

The proximal and distal regions of the ECR differed significantly in terms of the percentage of fibers classified as type I based on mATPase and myosin immunocytochemistry (fig. 1, table 1). Proximal muscle samples contained a mean of 12.2% type I fibers compared to a mean of 43.8% in the distal samples of the muscle. Similarly, there was a

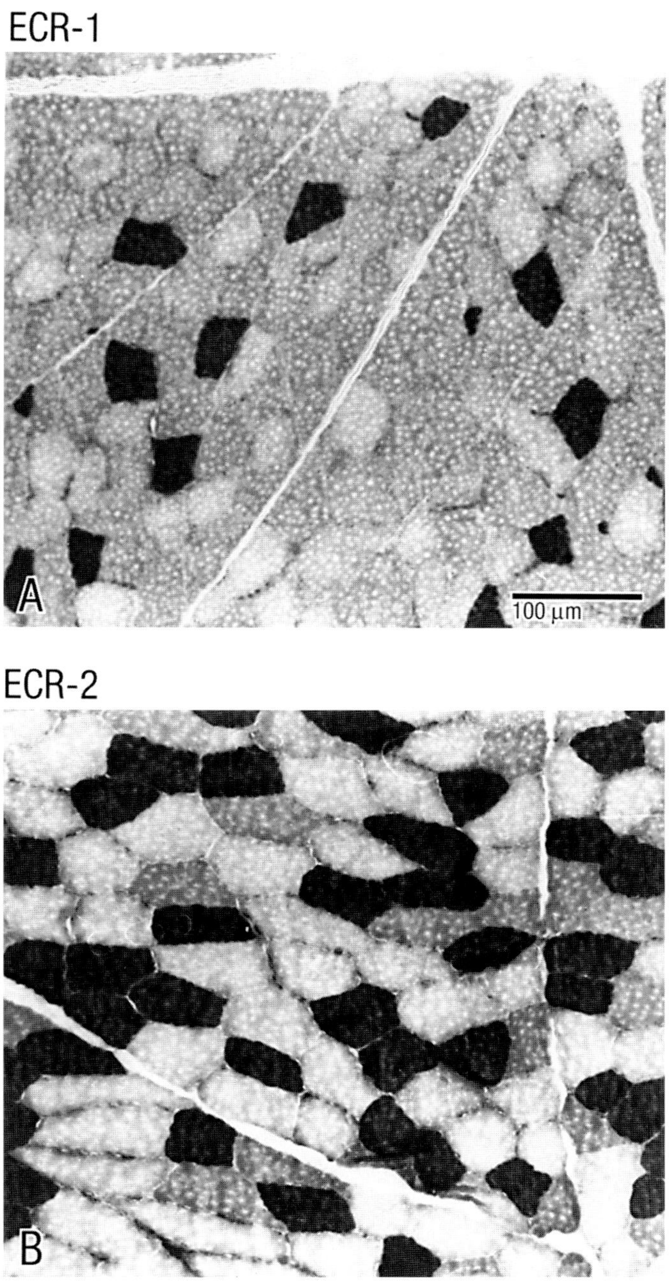

Fig. 1. Transverse sections of horse ECR muscle stained for myosin ATPase activity at pH 9.4, after preincubation at pH 4.45. Darkest fibers are classified as type I, presumed slow-twitch. The intermediate fibers are histochemical type IIb fibers [although electrophoretic considerations suggest classification as IIb/x: see Rivero et al., 1996]. Lightest fibers are type IIa. Both categories of type II fibers are presumed fast-twitch. Type I fibers are sparsely represented in the proximal region of the ECR muscle (**A**) whereas nearly 50% of the fibers seen in the distal region (**B**) are type I.

Table 2. Mean fiber diameters (μm) of fiber types in ECR muscles of horses

Horse	ECR (proximal)			ECR (distal)		
	I	IIa	IIb	I	IIa	IIb
1	42	50	63	57	67	–
2	44	54	69	56	60	50
3	27	43	58	40	50	51
4	33	46	56	46	56	56
5	40	44	53	47	57	55
6	59	70	73	45	63	66
7	20	39	55	34	59	51
Mean	38	50	61	47	59	57
±SE	5	4	3	2	2	2

reversal of the proportion of type IIb fibers with a mean of 62.5% in the proximal region and a relatively smaller mean of 22.5% in the distal samples. One horse, a 20-year-old thoroughbred, lacked IIb fibers in the distal region [demonstrated with histochemistry, immunocytochemistry, and electrophoresis of myosin heavy chain (MyHC) isoforms]. The distribution of type IIa fibers was more uniform along the proximal to distal axis, with mean percentages of 25.2 and 33.7, respectively.

The fiber type classifications were relatively straightforward. Type I fibers correlated with metabolic profiles of high oxidative and low glycolytic potential based on reaction intensity with NADH-TR and α-GPD reactions, respectively (see fig. 2). The type I fibers reacted positively against the antislow antibodies (S58 and NCL-MHCs) and did not react against antifast antibodies (MY32 and NCL-MHCf). The type IIa fibers correlated with relatively high oxidative and intermediate to high glycolytic potential, whereas the type IIb fibers were consistently highly glycolytic and nonoxidative. All type II fibers reacted strongly against antifast antibodies (both MY32 and NCL-MHCf), but did not react against the antislow antibodies.

Fiber Type Size Profiles

The type I fibers tended to be smallest, in the proximal compartment and, in 3 of the horses, exhibited many fibers with small profiles, perhaps representing tapered ends of fibers, or being smaller in general. The former seemed more plausible because the 'small fibers' tended to have a rather pinched appearance compared to the normal 'polygonal appearance' of normal muscle frequently described by myologists. The appearance of these relatively small fibers in 3 out of 7 horses may represent slight differences in sampling sites, or may be representative of interindividual differences. In the proximal compartment, the type IIb fibers were larger than the adjacent populations of type I and IIa fibers. In contrast, the type IIa and IIb fibers were not different in the distal compartment where the IIa fibers had a larger mean diameter than did the IIb fibers. The type I fibers were the smallest fibers in the distal compartment (mean 47 μm), but were larger than the type I fibers (mean 38 μm) observed in the proximal compartment.

Fiber Architecture

Fiber bundles in the proximal region of the muscle were approximately 92 mm in length whereas fiber bundles in the distal sample, adjacent to the site of histochemical study, were about 31 mm long. This estimate was based on measurements of 10 bundles obtained from tissue adjacent to the histochemical sample site. It is critical to know whether fibers in a bundle uniformly span the length of the bundle, or whether some population of the fibers may be shorter than others and terminate intrafascicularly. Fibers tended to span the entire length of the short bundles (31 mm length) studied or were approximately 2–4 mm shorter than the bundle (based on a measurement of 10 fibers from within each of the bundles reported above). Because of the brittle nature of the fibers in the long bundles (92 mm length), it was not possible to quantitate fiber lengths in the proximal compartment. In summary, the predominantly faster fibers (assuming that type II fibers are faster than type I fibers in the horse) are concentrated in the proximal portion of the muscle.

Fiber architecture studies on both intact and acid-digested specimens suggest that there may be three neuromuscular compartments. There is a long-fibered proximal compartment (sampled histochemically) with fiber origins along the lateral epicondyle of the humerus and insertions along the dorsal/distal connective tissue sheath. This connective tissue sheath continues distally to fuse with the tendon of insertion. A distal, short-fibered compartment (also sampled histochemically) contains fibers that originate along the connective tissue sheath on the deep surface of the muscle, and insert along the broad tendon of insertion in the distal one third of the muscle belly. A third compartment (not studied histochemically) shows relatively short fiber bundles (mean of about 45 mm) that originate along the lateral and deep surface of the muscle extending from the proximal one quarter to the midpoint of the muscle belly, and insert on a tendinous inscription that runs deep within the muscle along its middle two quarters. The latter tendinous inscription fuses distally with the ECR tendon of insertion.

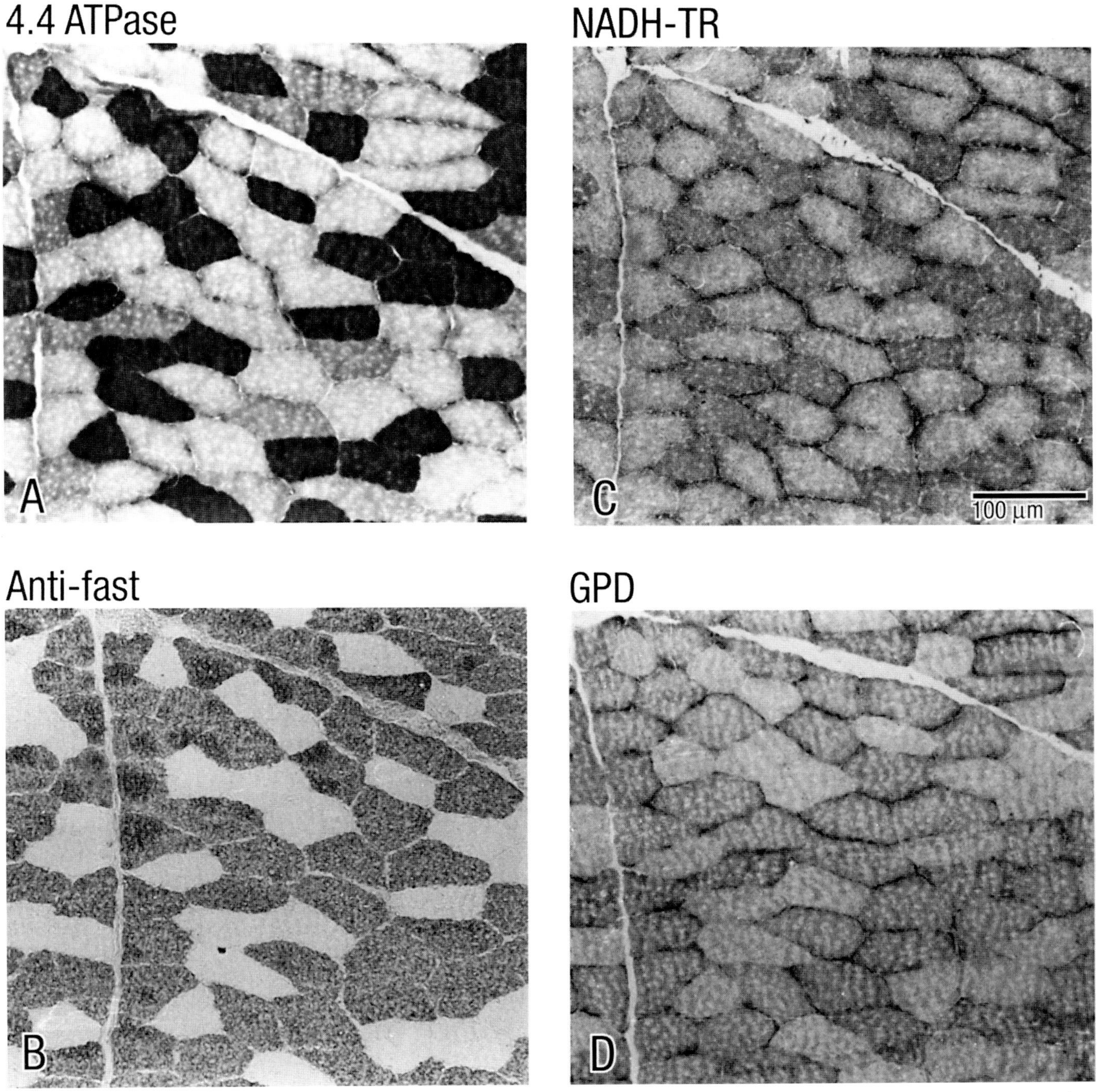

Fig. 2. Serial transverse sections of the distal compartment of equine ECR muscle stained for myosin ATPase activity after preincubation at pH 4.45 (**A**), reacted against antifast (MY32) antibodies (**B**), or stained for NADH-TR (**C**) or α-GPD (**D**). In general, type I fibers stained positively for NADH-TR, and negatively for α-GPD, and the antifast antibody. Type IIa fibers stained intensely for NADH-TR and for α-GPD, and reacted strongly against antifast antibodies. Type IIb/x fibers stained intensely for α-GPD, but not for NADH-TR, and they reacted strongly against antifast antibodies.

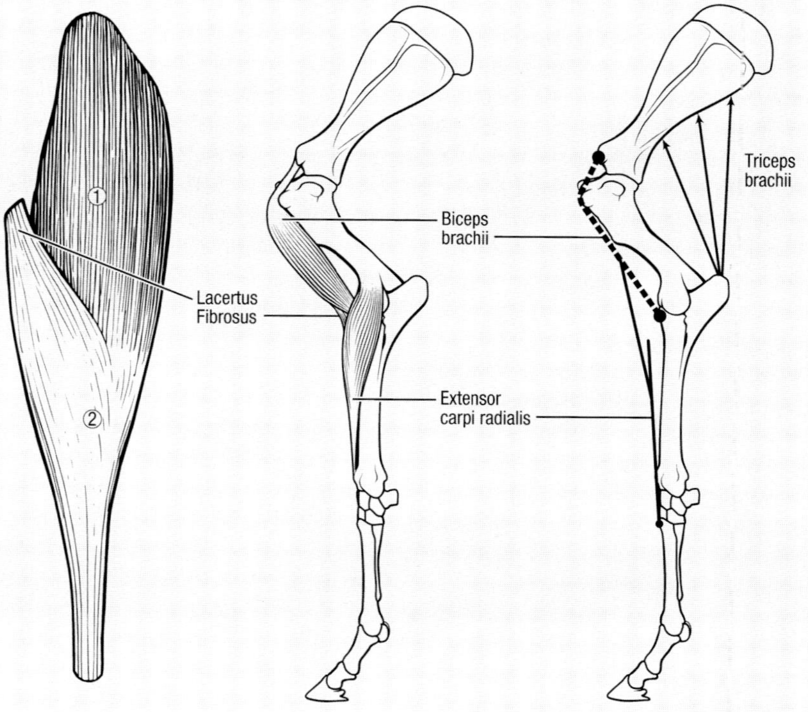

Fig. 3. Schematic representations of the ECR muscle of horses. Illustrated at the left is a cranial view of the ECR muscle showing the position of insertion of the lacertus fibrosus from the biceps brachii muscle (medially on the ECR). Long fibers are visible in the proximal region after stripping off the relatively thin surface connective tissues. In contrast, a thick band of connective tissue is distributed across the surface of the muscle distal to the insertion of lacertus fibrosus. This connective tissue band is continuous distally with the tendon of insertion of the ECR. Proximal and distal histochemical sample sites are indicated by the numbers 1 and 2, respectively. Illustrated in the middle panel is a lateral view of the equine forelimb showing the in-series relationship of the biceps brachii and the ECR. At the right, a simplified model highlights lines of action of the major intrinsic muscles of the upper forelimb and the ECR. Essentially, the line of action of the biceps brachii during passive stance pulls most directly on the distal, short-fibered compartment of the ECR. Fixation of the length of the biceps brachii which contains a noncompliant internal tendon extending from origin to insertion, will prevent collapse (flexion) of the shoulder under the influence of gravity. Similarly, 'fixation' of the tendon of insertion of the ECR by its fatigue-resistant short-fibered region will resist collapse (flexion) of the carpus region of the distal limb. Other presumed postural muscles in the forelimb include deep parts of the triceps brachii complex (anconeus and medial triceps, not shown) which are composed entirely of type I, fatigue-resistant fibers and are proposed to fix the elbow, preventing collapse during standing.

The lacertus fibrosus, a tendinous slip from the distal biceps brachii, inserts by fusing with the medial side of the perimysium of the ECR, about one third the distance from the muscle's origin. This coincides with the proximal extent of the tendon of insertion of the ECR. All of the 'long' fiber bundles of the ECR are situated proximal to the insertion point of the lacertus fibrosus (fig. 3). Fibers inserted on the insertional tendons at angles of approximately 45° along the distal half of the muscle. This was the region characterized by relatively shorter muscle fibers. Proximally, the angle of insertion tended to be less than 30° in a region containing the relatively longer fibers of the proximal compartment. The suggestion of three muscular compartments, based on fiber architecture, is reinforced by the observation in two ECR muscles obtained from 2 animals, of the distribution of three primary nerve branches to (1) the superficial and proximal part of the ECR (the proximal long-fibered region), (2) to the proximal and deep part of the ECR, and (3) to the deep surface of the distal one half of the muscle (the histochemically studied short fiber region).

Fig. 4. MyHC isoforms from horse ECR muscles identified by SDS-PAGE on 4% acrylamide gels. At the left is the two-isoform pattern expressed in the horse crural diaphragm, lacking the type IIb/x isoform. The proximal ECR contains three isoforms demonstrating increasing mobility on the gel, including type IIa, IIb/x, and type I, respectively. The distal ECR sample contained relatively more type I MyHC isoform than did the proximal sample. At the right is a sample obtained from an adult rat costal diaphragm to illustrate the 'standardized' positions of the type IIa, IIx, IIb and I MyHC isoforms.

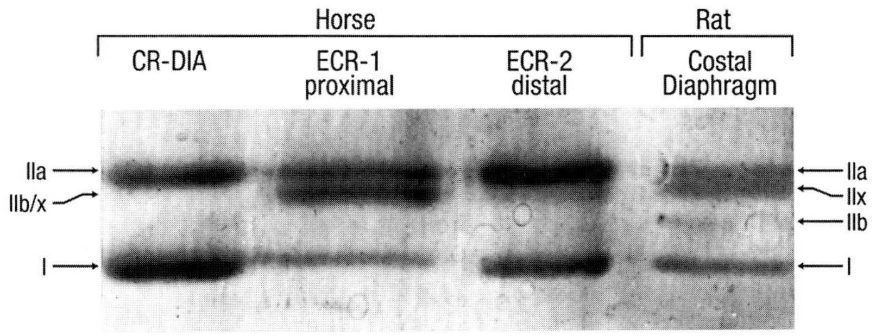

Electrophoretic Studies

Analysis of myosin heavy chain isoforms (MyHC) largely substantiated the findings based on histochemistry. In the equine ECR, there were three MyHCs, type I, IIb, and IIa, in order of decreasing mobility on the gels (see fig. 4). Based on the conclusions of Rivero et al. [1996], it is likely that the IIb fibers are best described as IIb/x since their mobility most closely matches the IIx isoform of rat diaphragm, and since they cross-reacted with antibodies to the MyHC types IIb and IIx. In general, the MyHC isoforms observed from the proximal samples of muscles expressed more intense silver staining in the IIa and IIb/x regions, whereas the MyHC isoforms taken from distal samples had relatively more intense silver staining associated with the type I MyHC. This pattern would be consistent with the fiber type proportions described by histochemical analysis.

Discussion

The concept of neuromuscular compartments is not new. The idea of a compartment was first discussed by English and Letbetter [1982a, b] and later refined by several studies from English's laboratory [English and Weeks, 1984; Gatesy and English, 1993]. Other groups noted similar designs in muscles from cats [Gonyea and Ericson, 1977; Botterman et al., 1978; Richmond et al., 1985; Edgerton et al., 1987; Bodine et al., 1989] and dogs [Latorre et al., 1993]. A parallel literature exists with respect to some of the muscles of mastication [see Rowlerson et al., 1988; Herring, 1992] as well as the complex axial muscles that are affiliated with the vertebral column [see Richmond et al., 1985; Armstrong et al., 1988; Selbie et al., 1993].

The present study suggests there are at least two, perhaps three compartments in the ECR of the horse and extends on observations made on other muscles of the horse [see biceps brachii: Hermanson and Hurley, 1990; deep digital flexor muscle: Hermanson and Cobb, 1991] which seem to exhibit design features blending postural and dynamic adaptations.

Burkholder et al. [1994] recently published an elegant model of appendicular muscles in the mouse based on measurements of whole muscle length, mass, fiber length, architecture (fiber orientation or pennation), and histochemical properties. They demonstrated a discrimination among the hindlimb muscles that appeared to emphasize the primacy of fiber architecture over histochemical patterns. For example, the larger, strap-like muscles of the caudal thigh (hamstrings) tended to have long, fast-twitch parallel fibers capable of producing high velocity shortening. In contrast, they noted that most of the crural muscles were packaged with shorter fibers, higher degrees of pinnation, and a tendency to emphasize oxidative (fatigue-resistant) properties. An earlier paper by Armstrong and Phelps [1984] documented the general histochemical (and proposed functional) properties of all hindlimb muscles in the rat. The smaller size of the rat and mouse muscles, relative to horse muscles, permits quantification of fiber type properties based on one or few transverse histochemical sections. Similarly, their small size and apparently reduced complexity permit measurement of pennation angle. One must remain aware, however, that many compartments are conserved in other species despite large size and phylogenetic differences. For example, Brasseur et al. [1987] noted predictable regional fiber type patterns in the medial gastrocnemius of laboratory mice despite the lack of discrete connective tissue divisions that might highlight individual 'compartments'. Donahue and English [1989] documented similar neuromuscular compartments in hindlimb muscles of rats as had been described in cats [English and Letbetter,

1982a]. These papers have proven inspirational to students following in this area of muscle biology. It appears, however, that the complexity of many of the larger muscles seen in the forelimb (and hindlimb) of horses may preclude a similar approach to the analysis and modelling of muscle function in horses. Thus, groups studying horse muscles have tended to focus on one or two sample sites to form their conclusions [see Rivero et al., 1996; Serrano et al., 1996, for review of relevant citations]. Caution is necessary to avoid exclusion of functionally important subdivisions within a large muscle that may be obscured by heavy connective tissue or that lie deep within the muscle.

The ECR of the horse is significant because of its relatively large size in the forearm extensor compartment, and because of its presumed role in facilitating stance. The two-burst EMG pattern described for ECR during walking in the horse [Jansen et al., 1992] suggests potential for complicated activity patterns during locomotion. The three-compartment architecture described in the present study only reinforces the need for caution in deciding about a function for this muscle. It is not possible to say that all fibers within a single compartment participate equally in any specific function either. For example, my preliminary study demonstrates a gradual shortening of mean fiber bundle length, in concert with an increasing angle of insertion on the distal tendon, of fibers in the proximal/superficial compartment. It is beyond the scope of this paper to address the architecture of the entire muscle. A more conclusive result may be obtained by breaking the muscle down into smaller components, the compartments, and measuring fiber parameters within each compartment, and weighing each compartment to assess the relative contribution to the muscle's physiological cross-sectional area made by long fibers, or by short fibers, for example. For the present, the data were polarized by describing two ends of the muscle and the obvious architectural and histochemical differences within them.

Although one might apply the concept of symmorphosis, championed by Taylor and colleagues [see Hoppeler et al., 1981; Armstrong et al., 1992] to describe the muscles of the athletic horse, this approach may obscure patterns that are imposed by other behavioral characteristics of the animal. Symmorphosis defines an economy of all the parts such that muscles will have a minimum number of mitochondria, or an appropriate diffusion distance between the vascular supplies and the mitochondria of active muscle fibers, sufficient to permit the animal to function at a rate necessary for survival in nature. The latter might be modelled as $\dot{V}O_2$max. Horses, however, spend as much as 23 h of each day standing, often relatively motionless [Ruckebush, 1972] during sleep and in a metabolic state far removed from $\dot{V}O_2$max. Thus, modern horse limbs have design features including passive mechanisms, specifically nontiring mechanisms to facilitate stance [Hermanson and MacFadden, 1992, 1996; Dyce et al., 1996]. The ECR may reflect this sort of bicameral design such that the long fibers of the proximal compartment undergo significant excursions related to elbow flexion and extension. In contrast to the proximal fibers, distal fibers tend to be more oxidative, more fatigue-resistant, and are shorter, suggesting less potential for excursion changes. The situation may be confounded by the insertion of the lacertus fibrosus from biceps brachii upon the proximal third to middle of the ECR belly (fig. 3). The lateral head of biceps brachii contains primarily oxidative, fatigue-resistant fibers and has a role in maintaining the extension of the shoulder against the force of gravity [Hermanson and Hurley, 1990]. The two muscles could act in-series to link shoulder extension with extension at the carpus [an antigravity role in both cases: see Dyce et al., 1996]. This in-series arrangement of the two muscles is comparable to the two-compartment design of the semitendinosus muscle in many species [Bodine et al., 1982]. Physiological study of the cat semitendinosus has shown that when proximal and distal compartments are stimulated, the distal compartment may shorten relatively more than the proximal compartment [Edgerton et al., 1987], suggesting independence of motor control and passive biomechanical properties. It is not clear how well matched activity in the two compartments is during normal locomotion, or if activity patterns shift with gait changes. Certainly the interposition of the long-fibered proximal compartment of the horse ECR can modulate or perhaps overwhelm the ongoing activity in the lateral head biceps brachii-ECR postural system (fig. 3). Electromyographic and kinematic data are required from standing and moving horses to clarify the functional properties of these muscles.

Acknowledgements

I wish to thank Dr. Matt Cobb and Dr. Jaqueline Petrie for their expert assistance in collecting and processing tissues used in the preliminary phases of this study. I thank Marketa Lillard for providing excellent technical assistance in the laboratory, particularly with the electrophoretic studies. Dr. John Bertram provided a review and discussion of initial drafts of this manuscript. This work could not be performed without the cooperation of the Department of Pathology (Necropsy Service) at the College of Veterinary Medicine. This study wa funded largely by a grant from the H.M. Zweig Memorial Fund for Research.

References

Andrews, F.M., T.L. Spurgeon (1986) Histochemical staining characteristics of normal horse skeletal muscle. Am J Vet Res *47:* 1843–1852.

Armstrong, J.B., P.K. Rose, S. Vanner, G.J. Baker, F.J.R. Richmond (1988) Compartmentalization of motor units in the cat neck muscle, biventer cervicis. J Neurophysiol *60:* 30–45.

Armstrong, R.B., R.O. Phelps (1984) Muscle fiber type composition of the rat hindlimb. Am J Anat *171:* 259–272.

Armstrong, R.B., B. Essen-Gustavsson, H. Hoppeler, J.H. Jones, S.R. Kayar, M.H. Laughlin, A. Lindholm, K.E. Longworth, C.R. Taylor, E.R. Wiebel (1992) O_2 delivery at $\dot{V}O_2$max and oxidative capacity in muscles of standardbred horses. J Appl Physiol *73:* 2274–2282.

Bodine, S.C., R.R. Roy, D.A. Meadows, R.F. Zernicke, R.D. Sacks, M. Fournier, V.R. Edgerton (1982) Architectural, histochemical, and contractile characteristics of a unique biarticular muscle: The cat semitendinosus. J Neurophysiol *48:* 192–201.

Botterman, B.R., M.D. Binder, D.G. Stuart (1978) Functional anatomy of the association between motor units and muscle receptors. Am Zool *18:* 135–152.

Brasseur, J.E., R.L. Curtis, J.W. Mellender, A.A. Rimm, J.L. Melvin, A.R. Sulaiman (1987) Systematic distribution of muscle fiber types in the medial gastrocnemius of the laboratory mouse: A morphometric analysis. Anat Rec *218:* 396–401.

Bruce, V., R. Turek (1985) Muscle fibre variation in the gluteus medius of the horse. Equine Vet J *17:* 317–321.

Burkholder, T.J., B. Fingado, S. Baron, R.L. Lieber (1994) Relationship between muscle fiber types and sizes and muscle architectural properties in the mouse hindlimb. J Morphol *221:* 177–190.

Dyce, K.M., W.O. Sack, C.J.G. Wensing (1996) Textbook of Veterinary Anatomy, ed 2, Philadelphia, Saunders.

Edgerton, V.R., S.C. Bodine, R.R. Roy (1987) Muscle architecture and performance. Stress and strain relationships in a muscle with two compartments in series. Med Sport Sci *26:* 12–23.

English, A.W., W.D. Letbetter (1982a) A histochemical analysis of identified compartments of cat lateral gastrocnemius muscle. Anat Rec *204:* 123–130.

English, A.W., W.D. Letbetter (1982b) Anatomy and innervation patterns of cat lateral gastrocnemius and plantaris muscles. Am J Anat *164:* 67–77.

English, A.W., O.I. Weeks (1984) Compartmentalization of single muscle units in cat lateral gastrocnemius. Exp Brain Res *56:* 361–368.

Gatesy, S.M., A.W. English (1993) Evidence for compartmental identity in the development of the rat lateral gastrocnemius muscle. Dev Dyn *196:* 174–182.

Gonyea, W.J., G.C. Ericson (1977) Morphological and histochemical organization of the flexor carpi radialis muscle in the cat. Am J Anat *148:* 329–344.

Hermanson, J.W., M.A. Cobb (1991) The extensor carpi radialis muscle of horses: What is learned from muscle architecture? American Association of Veterinary Anatomists conference, Pullman.

Hermanson, J.W., M.A. Cobb (1992) Four forearm flexor muscles of the horse, *Equus caballus:* Anatomy and histochemistry. J Morphol *212:* 269–280.

Hermanson, J.W., M.T. Hegemann-Monachelli, M.J. Daood, W.A. LaFramboise (1991) Correlation of myosin isoforms with anatomical divisions in equine musculus biceps brachii. Acta Anat *141:* 369–376.

Hermanson, J.W., K.J. Hurley (1990) Architectural and histochemical analysis of the biceps brachii muscle of the horse. Acta Anat *137:* 146–156.

Hermanson, J.W., B.J. MacFadden (1992) Evolutionary and functional morphology of the shoulder region and stay-apparatus in fossil and extant horses. J Vertebr Paleontol *12:* 377–386.

Hermanson, J.W., B.J. MacFadden (1996) Evolutionary and functional morphology of the knee in fossil and extant horses (Equidae). J Vertebr Paleontol *16:* 349–357.

Herring, S.W. (1992) Muscles of mastication: Architecture and functional organization; in Davidovitch Z (ed): The Biological Mechanisms of Tooth Movement and Craniofacial Adaptation. Columbus, Ohio State University, pp 541–548.

Hoppeler, H., O. Mathieu, E.R. Weibel, R. Krauer, S.L. Lindstedt, C.R. Taylor (1981) Design of the mammalian respiratory system. VIII. Capillaries in skeletal muscles. Respir Physiol *44:* 129–150.

Jansen, M.O., J.A.G.M. van Raaij, A.J. van den Bogert, H.C. Schamhardt, W. Hartman (1992) Quantitative analysis of computer-averaged electromyographic profiles of intrinsic limb muscles in ponies at the walk. Am J Vet Res *53:* 2343–2349.

Latorre, R., F. Gil, J.M. Vázquez, F. Moreno, F. Mascarello, G. Ramirez (1993) Morphological and histochemical characteristics of muscle fibre types in the flexor carpi radialis of the dog. J Anat *182:* 313–320.

Novikoff, A.B., W. Shin, J. Drucher (1961) Mitochondrial localization of oxidative enzymes: Staining results with two tetrazolium salts. J Biophys Biochem Cytol *9:* 47–61.

Richmond, F.J.R., D.R.R. McGillis, D.A. Scott (1985) Muscle fiber compartmentalization in cat splenius muscle. J Neurophysiol *53:* 868–885.

Rivero, J.-L., A.L. Serrano, P. Henckel, E. Agüera (1993) Muscle fiber type composition and fiber size in successfully and unsuccessfully endurance-raced horses. J Appl Physiol *75:* 1758–1766.

Rivero, J.-L., R.J. Talmadge, V.R. Edgerton (1996) Myosin isoforms in adult equine skeletal muscle: An immunohistochemical and electrophoretic study. Anat Rec *246:* 185–194.

Rowlerson, A., F. Mascarello, D. Barker, H. Saed (1988) Muscle-spindle distribution in relation to the fibre-type composition of masseter in mammals. J Anat *161:* 37–60.

Ruckebush, Y. (1972) The relevance of drowsiness in the circadian rhythm of farm animals. Anim Behav *20:* 637–643.

Ryan, J.M., M.A. Cobb, J.W. Hermanson (1992) Elbow extensor muscles of the horse: Postural and dynamic implications. Acta Anat *144:* 71–79.

Selbie, W.S., D.B. Thomson, F.J.R. Richmond (1993) Suboccipital muscles in the cat neck: Morphometry and histochemistry of the rectus capitis muscle complex. J Morphol *216:* 47–63.

Serrano, A.L., J.L. Petrie, J.-L. Rivero, J.W. Hermanson (1996) Myosin isoforms and muscle fiber characteristics in equine gluteus medius muscle. Anat Rec *244:* 444–451.

Wattenberg, L.W., J.L. Leong (1969) Effects of coenzyme Q_{10} and menadione on succinate dehydrogenase activity as measured by tetrazolium salt reduction. J Histochem Cytochem *8:* 296–303.

Acta Anatomica

Acta Anat 1997;159:136–146

R.R. Roy
J.A. Kim
R.J. Monti
H. Zhong
V.R. Edgerton

Brain Research Institute and Physiological Science Department, University of California, Los Angeles, Los Angeles, Calif., USA

Architectural and Histochemical Properties of Cat Hip 'Cuff' Muscles

Key Words
Skeletal muscle architecture
Fiber-type composition
Compartmentalization
Hip muscles
Cat

Abstract
The architectural properties and fiber-type composition of the cat hip 'cuff' muscles, i.e. the deep layer of muscles encircling the hip joint, were determined. The muscles studied included the gluteus minimus, obturator internus and externus, gemellus superior and inferior, pyriformis, quadratus femoris and capsularis. In addition, the fiber-type composition was determined for the iliacus muscle. Compared to other lower limb muscles, the cuff muscles were small (~0.2 to 2.7 g) and short (~15 to 40 mm long) and had short mean fiber lengths (~7 to 12 mm long), small angles of fiber pinnation (≤12°), and small physiological cross-sectional areas (~0.2 to 2.8 cm^2). The percentage of the cross-sectional area comprised of slow fibers ranged from 24 to 95% with this value being over 50% in 5/9 muscles studied. The small angle of pinnation and short fiber lengths optimize force production and the relatively high percentage of slow fibers suggest a high level of activation. Both of these properties are consistent with a hip stabilization role for these muscles. In addition, the high percentage of slow fibers suggests a high spindle density and a possible role of these muscles in providing proprioceptive feedback to the central nervous system for the control of posture and locomotion.

Introduction

The fiber-type composition of a skeletal muscle is an important indicator of its functional behavior. Muscles or muscle regions comprised of a large percentage of slow oxidative (SO) fibers are associated with relatively high activation rates even during slow movements and postural activity [Pierotti et al., 1989; Roy et al., 1991]. In contrast, muscles or muscle regions with a high percentage of fast glycolytic (FG) fibers appear to be recruited primarily during high intensity activity.

Similarly, the architectural design of a skeletal muscle is an important determinant of its functional capabilities [Gans and Bock, 1965; Close, 1972; Josephson, 1975; Spector et al., 1980; Bodine et al., 1982; Gans, 1982; Muhl, 1982; Sacks and Roy, 1982; Edgerton et al., 1983; Powell et al., 1984; Wickiewicz et al., 1984; Gans and de Vree, 1987; Otten, 1988; Lieber and Blevins, 1989; Peters, 1989; Roy and Edgerton, 1992; Burkholder et al., 1994]. For example, muscles with short fibers, small fiber length:muscle length ratios and/or large angles of fiber pinnation are optimally designed to maximize force production, whereas muscles

KARGER
Fax +41 61 306 12 34
E-Mail karger@karger.ch
www.karger.com

Roland R. Roy, PhD
Brain Research Institute, UCLA School of Medicine, Center for the Health Sciences
10833 Le Conte Avenue, Los Angeles, CA 90095-1761 (USA)
Tel. +1 310 825-2193, Fax +1 310 206-9184
E-Mail rrr@ucla.edu

having long fibers, large fiber length:muscle length ratios and/or minimal angles of fiber pinnation are designed to emphasize velocity of movement.

The purpose of the present study was to determine the architectural and fiber type distribution of the deep layer of muscles in close contact with and functioning at the hip joint in the cat. Because of their location, we assumed that these muscles function as hip stabilizers. Our basic hypothesis was that these stabilizing muscles would have a relatively high percentage of SO and/or fast oxidative glycolytic (FOG) fibers and would have relatively short fiber lengths, features which emphasize force production over prolonged periods. Preliminary data have been published in abstract form [Roy et al., 1983].

Materials and Methods

Adult male and female cats used for other studies not involving the hip musculature (2.4–4.7 kg, body weight) were overdosed with sodium pentobarbital. The hindlimbs were prepared for architectural determinations as described by Sacks and Roy [1982] or for fiber-type determination using the procedures described by Peter et al. [1972]. Nine muscles which surround the hip joint were studied: iliacus, gluteus minimus, obturator internus and externus, gemellus superior and inferior, pyriformis, quadratus femoris and capsularis. The iliacus was very difficult to separate from the psoas major; therefore, only a general description is provided for this muscle. All procedures used in this study were approved by the Animal Use Committee at UCLA and followed the American Physiological Society Animal Care Guidelines.

Architectural Determinations

Briefly, the trunk and hindlimbs were skinned and the trunk transected at ~T12 and the limbs cut at midfemur. The connective tissue and the superficial muscles around the hip region were removed to completely expose the deep layer of muscles. Intermuscular fascial sheaths were separated to allow for maximum exposure of the deep muscles. The specimen then was pinned securely to a stiff styrofoam board with the hip at ~100 to 110° from full extension and slight abduction. The specimen was then submerged in 10% buffered formalin for a minimum of 1 week. Individual muscles were removed carefully and the length of each muscle was measured, i.e. the distance between the termination of the muscle fibers at the origin and insertion, using a vernier caliper. The muscles were laid flat on a glass plate that was illuminated from below and the base of a protractor was oriented along the estimated line of pull through the origin and insertion. A string secured at the origin of the protractor scale was then adjusted until it paralleled the muscle fascicles. An approximate angle of pinnation of the fibers in one dimension was measured in 2–5 regions depending on the fiber arrangement. Variations in the angle of pinnation were negligible in most muscles. In these cases the mean of these values was taken as an estimate of the pinnation angle in one dimension. In some muscles the fiber arrangement was more complicated and these are described in 'Results'. The muscles were placed in a 0.4 *M* phosphate-buffered solution, pH 7.2, for 1–2 days and then transferred to a 15–20% sulfuric acid solution to allow maceration of the connective tissue. When bundles of fibers could easily be teased apart (ranging from 2 to 8 days), the muscle was rinsed overnight in the phosphate solution and then placed in a 50% glycerol solution for storage. Fiber bundles containing 5–10 fibers were dissected from several regions of the muscle using fine forceps and a binocular dissection scope (×80) and individual fibers from selected bundles were isolated and measured. The individual fibers were mounted on slides in glycerine jelly. All of these procedures were performed at room temperature. Using a light microscope (×400) and an eyepiece micrometer, sarcomere lengths were measured at three sites along the length of each fiber and a mean sarcomere length calculated. The length of the fibers were subsequently normalized to a sarcomere length of 2.2 μm [Gordon et al., 1966] to account for any variability that might result from the angle of joint fixation. These architectural values then were used to calculate the physiological cross-sectional area (PCSA) of each muscle as follows:

PCSA = (muscle wet weight) (cosine theta)/(mean adjusted fiber length) (muscle density)

where the muscle mass was the wet weight of the muscle (g); theta was equal to the mean approximate angle of fiber pinnation in one dimension (°); mean fiber length was normalized to a sarcomere length of 2.2 μm (cm); and muscle density was assumed to be 1.056 g/cm^3 [Mendez and Keys, 1960].

Histochemical Procedures

Individual muscles were removed, cleaned of excess fat and connective tissue, gently stretched to approximate their in vivo length, pinned to cork and quick frozen by immersion in isopentane cooled in liquid nitrogen. Subsequently, each muscle was divided into blocks in a cryostat using a sharp, cold steel blade such that cross sections could be cut from the center of the proximal, middle and distal one-third of the muscle. Each muscle block was then placed on cork with the fibers oriented perpendicular to the surface of the cork.

Transverse sections (10 μm thick) were cut from each frozen tissue block using a rotary cryostat at –20 °C. Serial sections were stained for the following contractile and metabolic enzymes: myosin adenosine triphosphatase (myofibrillar ATPase with alkaline or acid preincubation) [Nwoye et al., 1982], α-glycerophosphate dehydrogenase [Wattenberg and Leong, 1960] and succinic dehydrogenase [Barka and Anderson, 1963]. To determine muscle fiber types, muscle transverse sections then were projected on a wall using a microprojector. This procedure allowed for the entire muscle cross section stained for myofibrillar ATPase (alkaline preincubation) to be displayed and for the selection of representative samples of muscle fibers. A sample of ~200 fibers was identified from 2 or 4 regions of each muscle cross section for a total sample of 400–800 fibers from each muscle level (table 2). This sampling procedure provides at best an approximation of the fiber type composition of each muscle, particularly for those muscles showing a high degree of regionalization. Anatomical landmarks such as connective tissue septa and large blood vessels and nerves were used to ensure a reasonable consistency in selecting the sampling sites within and across cats. The fiber type distributions across all three levels of the quadratus femoris and obturator externus are shown in figure 1. A light microscope was used to locate the same muscle fibers in serial sections stained for myofibrillar ATPase (acid preincubation) and the metabolic enzymes. Individual fibers were classified as fast glycolytic (FG), fast oxidative glycolytic (FOG) or slow oxidative (SO) based on the qualitative criteria established by Peter et al. [1972]. A small percentage of fibers (<2% in any single muscle) did not clearly fit the classification scheme in that the myosin ATPase reaction did not completely reverse when staining in an acid or alkaline prein-

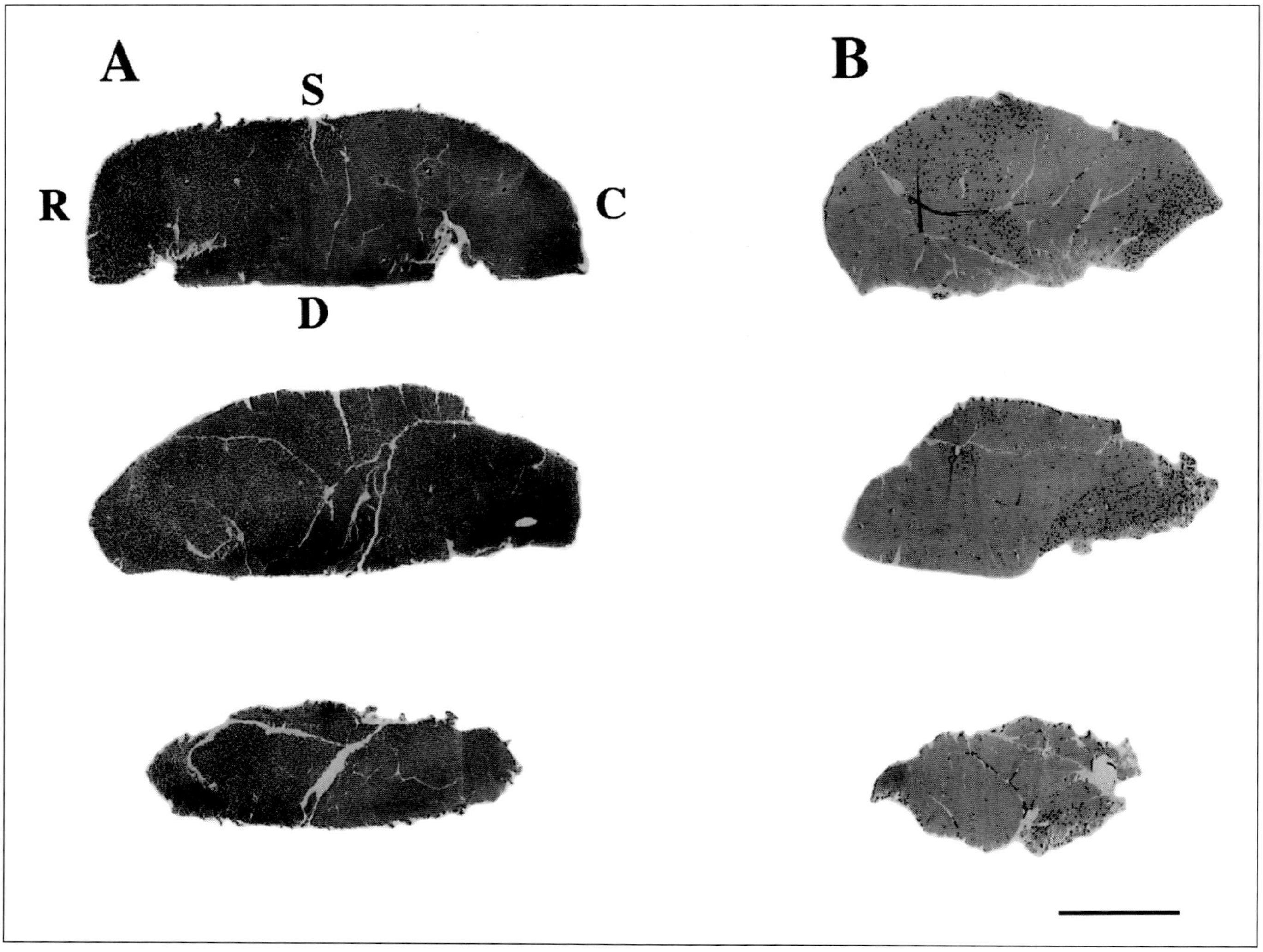

Fig. 1. Fiber-type composition of the obturator externus (**A**) and quadratus femoris (**B**) at the proximal (top), middle and distal (bottom) thirds of the muscles. Myofibrillar ATPase (alkaline preincubation) stain with light staining fibers being slow. Orientation in **B** is identical to that in **A**. R = Rostral; S = superficial (away from the bone); C = caudal; D = deep (close to the bone). Bar = 5 mm.

cubation medium; these fibers were excluded from further analyses in this study. An overall estimate of the relative contribution of each fiber type to the total cross-sectional area of the muscle was determined for the middle block of each muscle as follows. A planimeter was used to outline the cross-sectional area (region) represented by each fiber sample and a percent cross-sectional area calculated. This percent was then used to weight the fiber type composition of each sample and an overall mean was determined. All data are presented as means ± SEM.

Results

General Comments

The actions of the muscles were taken from Crouch [1969] and, in some cases, were modified based on the gross manipulation of the limbs via pulling the tendons during dissection. In each muscle, many fibers appeared to run the entire distance between the tendinous planes of origin and insertion to terminate at each end in a blunt type of ending [Trotter, 1993]. Some fibers, however, terminated intrafascicularly and appeared to taper in the manner described previously [Loeb et al., 1987; Richmond and Armstrong, 1988; Trotter, 1991; Ounjian et al., 1991; Eldred et al., 1993]. Some of the muscles had a complex architectural arrangement, apparently being highly compartmentalized anatomically. This was apparent both from gross dissections and from the arrangement of fiber types within individual muscles. In the present study, however, we did not attempt to identify individual compartments via the stimulation of primary nerve branches as described previously [English and Letbetter, 1982a, b; English and Weeks, 1987]. Some gross descriptions of this compartmentalization will be given for individual muscles.

Although there was animal-to-animal variability, the general trend in fiber type composition was similar across cats. This similarity was observed for comparisons across regions within a cross-section and within regions across muscle levels (fig. 1). Connective tissue septa were used to follow regions across the levels of the muscle. The architectural and fiber type data are summarized in tables 1 and 2, respectively.

Capsularis

The capsularis presumably is a weak flexor and medial rotator at the hip joint. However, its location and high spindle density also suggests that it serves as a 'sensor' muscle [Eldred et al., 1997]. The capsularis muscle originates on the lateral surface of the ilium in front of the acetabulum and inserts onto the anterolateral aspect of the superior end of the femur. The lateral (actually facing caudally, laterally and inferiorly) border of the capsularis is ~2–3 mm longer than the medial border.

The mean wet weight of the muscle was 236 mg and the average muscle length was 25.4 mm. Fiber lengths ranged from ~8 to 16 mm. Thus, the fiber length:muscle length ratios ranged from 0.31 to 0.63 with an overall average of 0.47. Fibers tended to be longer near the lateral than the medial border of the muscle, i.e. the region that during extension of the hip (in situ) is seen to undergo the greatest lengthening [Eldred et al., 1997]. The fibers were arranged in parallel and deviated less than 5° from the line of force through the bony attachments. Based on these architectural measurements, the average calculated PCSA was ~0.187 cm^2, the smallest of any of the 'cuff' muscles.

The capsularis was comprised predominantly of slow fibers and exclusively of oxidative fibers. There was a slightly higher percentage of FOG fibers in the superficial compared to the deep region and the caudal compared to the rostral region of the muscle. For each region studied, the fiber type composition was similar at all muscle levels.

Quadratus femoris

The quadratus femoris is an extensor and lateral rotator at the hip joint. The quadratus femoris originates by fleshy fibers from a broad region on the lateral surface of the ischium near the ischial tuberosity and inserts on the posterior border of the greater and lesser trochanters.

The mean weight of the quadratus femoris was 1.9 g and the mean length was 27.5 mm. There was a range in fiber lengths from 6 to 14 mm, with the deepest (near the bone) fibers being the shortest. The mean angle of fiber pinnation was ~12° and the mean fiber length:muscle length ratio was 0.36. The quadratus femoris had a relatively large PCSA of 1.795 cm^2.

The quadratus femoris was the slowest of all 'cuff' muscles, averaging 90% or higher SO fibers in all regions (see fig. 1B). The highest percentage (~10%) of FOG fibers was found in the caudal region of the muscle. The fiber type composition for each region was consistent at the three muscle levels studied.

Gluteus minimus

The gluteus minimus is an abductor and lateral rotator at the hip joint. The gluteus minimus originates from the ventral half of the lateral surface of the ilium and inserts into the inferior border of the greater trochanter on the lateral surface of the femur.

The mean weight and length of the gluteus minimus was 1.09 g and 40.1 mm, respectively. A sheet of tendon lined the muscle on its superficial surface. The mean angle of pinnation was ~10°. The range of fiber lengths was relatively small (8–13 mm), with the fibers in the superficial region being somewhat longer than the fibers in the deep region. An internal tendon ran the length of muscle on which the fibers attached. The mean fiber length:muscle length ratio was 0.26 and the mean PCSA was 0.968 cm^2.

The superficial region of the muscle had a relatively equal mixture of the three fiber types at each muscle level. A similar fiber type distribution was observed in the deep region of the proximal and middle levels of the muscle, whereas the distal level showed a much higher percentage of FG and lower percentage of SO fibers.

Pyriformis

The pyriformis is a strong abductor and very weak extensor at the hip joint. The pyriformis originates from fleshy fibers from the tips of the transverse processes from the last sacral to the second caudal vertebrae and inserts via a flat tendon into the inside edge of the proximal border of the greater trochanter. The muscle lays on top of the gemellus superior and underneath the gluteus medius.

The pyriformis had a mean weight of 1.43 g and mean length of 29.3 mm. Muscle fiber lengths were longer in the superficial compared to the deep portion of the muscle. The muscle was easily split into sections and a small tendinous band that ran the length of the muscle fanned out to provide an attachment site for fibers that were on the average ~40% the length of the muscle. The muscle was quite flat with the fibers having a very small deviation from the line of pull (<5°). The mean PCSA of the pyriformis was 1.163 cm^2.

Table 1. The architectural properties of cat hip 'cuff' muscles

Muscle	Muscle mass g	Muscle length mm	Fiber lengths (range) mm	Normalized fiber length mm	Fiber length/ muscle length mean (range)	Pinnation angle degrees	Estimated PCSA cm^2
Capsularis	0.236±0.080	25.4±3.3	8–16	11.9±1.9	0.47 (0.31–0.63)	<5	0.187
Quadratus femoris	1.90 ±0.63	27.5±3.1	6–14	9.8±0.7	0.36 (0.22–0.51)	12	1.795
Gluteus minimus	1.09 ±0.26	40.1±4.5	8–13	10.5±0.5	0.26 (0.20–0.32)	10	0.968
Pyriformis	1.43 ±0.32	29.3±2.9	8–16	11.6±1.4	0.40 (0.27–0.55)	<5	1.163
Gemellus superior	0.787±0.223	24.2±3.2	5–10	7.2±0.5	0.30 (0.21)–0.41)	12	1.013
Gemellus inferior	0.597±0.139	15.6±1.1	5–10	7.2±1.2	0.46 (0.32–0.64)	8	0.778
Obturator externus	1.90 ±0.35	29.4±2.9	6–12	8.4±0.4	0.29 (0.20–0.41)	6	2.131
Obturator internus	2.69 ±0.55	27.1±4.8	5–15	8.8±0.4	0.32 (0.18–0.55)	12	2.832

Values are means±SEM from 3 to 5 cats. Normalized fiber lengths, adjusted to a sarcomere length of 2.2 μm. PCSA = Physiological cross-sectional area.

Table 2. Fiber-type composition (percentage) of the cat hip 'cuff' muscles

Muscle	Region	Deep/midcaudal[a]			Superficial/midrostral[b]			Caudal/medial[c]			Rostral/lateral[d]		
	level	FG	FOG	SO	FG	FOG	SO	FG	FOG	SO	FG	FOG	SO
Capsularis[c, d]	proximal	0	8±5	92±5	0	16±9	84±9	0	8±4	92±4	0	14±9	86±9
	middle	0	9±6	91±6	0	17±9	83±9	0	12±6	88±6	0	15±8	85±8
	distal	0	7±4	93±4	0	15±8	85±8	0	10±4	90±4	0	10±5	90±5
Quadratus femoris	proximal	0	2±1	98±1	0	5±2	95±2	0	10±2	90±2	0	2±2	98±2
	middle	0	1±1	99±1	0	6±2	94±2	0	11±2	89±2	0	3±2	97±2
	distal	0	1±1	99±1	0	6±2	94±2	0	9±2	91±2	0	2±2	98±2
Gluteus minimus	proximal	38±2	30±3	32±2	35±2	35±3	30±1	–	–	–	–	–	–
	middle	39±1	33±2	28±1	38±2	27±2	35±2	–	–	–	–	–	–
	distal	65±4	22±4	13±1	39±1	29±1	32±1	–	–	–	–	–	–
Pyriformis[a, b]	proximal	21±2	26±4	53±4	22±3	27±7	51±4	28±4	30±5	42±1	22±1	25±5	53±4
	middle	21±3	24±3	55±2	22±1	28±5	50±5	29±3	28±3	43±1	18±1	25±7	57±6
	distal	19±1	28±5	53±5	25±2	25±6	50±4	29±1	26±3	45±2	18±3	27±6	55±3
Gemellus superior	proximal	45±4	35±3	20±2	28±3	22±3	50±0	0	24±3	76±3	47±2	34±2	19±1
	middle	48±10	28±7	24±4	30±3	19±3	51±2	0	23±3	77±3	39±1	32±1	29±1
	distal	64±3	19±4	17±2	28±3	33±3	39±1	0	18±3	82±3	35±5	32±3	33±2
Gemellus inferior	proximal	26±6	27±3	47±9	20±2	15±3	65±5	36±2	25±1	39±1	0	33±6	67±6
	middle	25±5	31±2	44±7	18±3	19±2	63±6	32±2	27±3	41±1	0	30±3	70±3
	distal	33±3	30±1	37±4	22±2	20±1	58±1	35±4	26±3	39±1	0	27±5	73±5
Obturator externus	proximal	72±2	20±2	8±1	41±3	27±3	32±1	52±3	23±4	25±2	30±2	35±3	35±2
	middle	70±2	23±2	7±1	44±2	25±2	31±3	51±3	26±2	23±1	34±3	28±1	38±3
	distal	66±3	23±2	11±1	51±1	29±1	20±1	58±1	25±1	17±1	45±3	33±4	22±3
Obturator internus[a, b]	proximal	64±2	15±2	21±1	60±3	15±4	25±1	77±2	20±1	3±1	42±3	21±2	37±1
	middle	62±3	12±1	26±4	57±4	18±3	25±1	78±1	17±2	5±1	43±9	19±4	38±4
	distal	74±6	22±5	4±1	61±1	18±5	20±5	76±1	23±2	1±1	59±1	19±7	22±7

Values are means±SEM from 3–4 cats. The regions were deep, superficial, caudal and rostral unless otherwise specified: [a] midcaudal; [b] midrostral; [c] medial, and [d] lateral.

Overall, the pyriformis was comprised of ~50% SO and ~20% FG fibers. The medial region was slightly faster, having ~40% and 30% SO and FG fibers, respectively. The fiber type distribution was very consistent across muscle levels.

Gemellus superior

The gemellus superior is an abductor and medial rotator at the hip joint. The gemellus superior originates along the border of the greater sciatic notch and inserts by a strong tendon on the medial aspect of the proximal border of the greater trochanter.

The gemellus superior averaged 0.787 g in weight and 24.2 mm in length. The muscle could be separated grossly into a rostral and caudal portion, with some of the fibers in the caudal portion passing under the fibers in the rostral portion. The fibers range in length from ~5 to 10 mm with a mean of 7.2 mm. The shortest fibers were located in the deep portion of the muscle, i.e. the portion nearest the joint capsule. The fibers had an average angle of pinnation of ~12° and the estimated PCSA was 1.013 cm^2.

The gemellus superior had a very large area in the caudal region that was comprised of all oxidative fibers, with a predominance of SO fibers. The rostral region displayed a mixture of fiber types. The central region showed a deep to superficial gradient of fiber types, with the superficial region having the highest percentage of SO fibers.

Gemellus inferior

The gemellus inferior is an abductor and lateral rotator at the hip joint. The gemellus inferior originates along the border of the lesser sciatic notch and inserts into the distal tendon of the obturator internus and the joint capsule. The fibers in the midportion of the muscle were intricately attached to the tendon of the obturator internus.

The gemellus inferior was somewhat smaller and shorter than the gemellus superior, averaging 0.57 g and 15.6 mm. The average and range in fiber length were identical to the gemellus superior, i.e. mean of 7.2 mm and range from 5 to 10 mm. The shortest fibers were found in the portion of the muscle attached to the tendon of insertion of the obturator internus and in the deeper portions of the other regions. The mean fiber angle of pinnation was ~8°. The mean PCSA was 0.778 cm^2.

The gemellus inferior showed a progressive decrease in SO fibers (from ~70 to 40%) and a progressive increase in FG fibers (from 0 to ~35%) moving from rostral to caudal. Generally, the fiber type composition was consistent across the three muscle levels.

Obturator externus

The obturator externus is an extensor and outward rotator at the hip joint. The obturator externus has a wide origin: from fleshy fibers from the medial lip of the obturator foramen, external surface of the rami of the pubis and ischium adjacent to the lip and from the outer surface of the ramus of the ischium as far as the area for the quadratus femoris. The muscle inserts on the bottom of the trochanteric fossa on the medial aspect of the greater trochanter.

The mean weight of the obturator externus was 1.89 g and the mean length was 29.4 mm. Fiber lengths ranged from 6 to 12 mm, with an average length of 8.4 mm and a mean fiber length:muscle length ratio of 0.29. The fiber lengths were slightly graded with the fibers nearest the joint capsule and in the center of the muscle being the shortest fibers. The angle of fiber pinnation was ~6°. The PCSA was relatively large, i.e. 2.131 cm^2.

The fiber type distribution of the obturator externus was highly regionalized (fig. 1A). The superficial portion of the muscle had a higher percentage of SO fibers than the deep region. In addition, the rostral portion of the muscle had a higher proportion of SO fibers than the caudal portion. The caudal portion was very fast, having 10% or less SO fibers at all muscle levels. Except for the fast deep region, the muscle had a higher percentage of fast fibers, primarily FG fibers, at the distal end compared to the middle and proximal levels.

Obturator internus

The obturator internus is an abductor and outward rotator at the hip joint. The obturator internus originates by fleshy fibers from the internal surface of the pubic and ischial rami from the symphysis pubis nearly to the ischial tuberosity and inserts on the bottom of the trochanteric fossa on the medial aspect of the greater trochanter. The tendon of insertion is unusally thick and strong and has a permanent crease (bend) in it due to its passage around the ischium to its insertion. This tendon also serves as the insertion of the mid-portion of the gemellus inferior muscle.

The obturator internus was the largest of the 'cuff' muscles studied, weighing 2.69 g. The mean muscle and fiber lengths were 27.1 and 8.8 mm, respectively, resulting in a fiber length:muscle length ratio of 0.32. The mean fiber angle of pinnation was ~15°. The obturator internus was a very complex muscle having a number of subdivisions and fiber pinnation arrangements. For example, at least 9 distal tendons inserted into a common tendon and each tendon was associated with a few large bundles of fibers. This muscle had the widest range in fiber lengths observed among the 'cuff' muscles and the fiber lengths were variable throughout the muscle. The angle of pinnation varied from bundle to bundle and was greater in the more central than in the more peripheral regions. The obturator internus had the largest PCSA of any muscle studied, i.e. 2.832 cm^2.

The obturator internus was comprised of the highest percentage of fast fibers of any hip 'cuff' muscle. For example, the caudal region averaged ~77% FG and only 3% SO fibers across the three muscle levels. In addition, both the midcaudal and midrostral regions had ~60–70% FG fibers. There also was an increase in the percentage of fast fibers in the distal compared to the proximal and middle portions of the muscle.

Iliacus

The iliacus, along with the psoas major, is a powerful flexor at the hip joint. The iliacus originates from the inside surface of the ilium from the sacroiliac joint to the iliopectineal eminence and inserts, in close contact with the psoas major, onto the lesser trochanter of the femur. The iliacus was intimately associated with the psoas major. For the purposes of the present study, the iliacus was separated from the psoas major by identifiying its origin and carefully separating the muscles to their common insertion. Unfortunately, this was not always a clean separation and, therefore, the muscle weights and lengths were not very accurate (data not included in table 1).

There was a central and deep core of fibers that were followed from their insertion on the trochanteric fossa to their origin on the ilium and these fibers were considered to be representative of the iliacus. There was a gradation of fiber lengths with the fibers in the deep region being shorter than the fibers in the superficial region of the muscle. In two muscles, fiber lengths range from ~14 to 28 mm in length, with a majority of the fibers showing long tapers at one end. In addition, a few fibers showed tapering at both ends. The angle of pinnation was minimal, i.e. less than 5°.

The iliacus appeared to be divided into a relatively slow and a relatively fast region and this arrangement was consistent along the length of the muscle. Across all three muscle levels studied, the lateral portion of the muscle was comprised of 73 ± 2, 25 ± 1 and 2 ± 2% and the medial portion 16 ± 1, 26 ± 2 and 58 ± 3% SO, FOG and FG fibers, respectively (data not included in table 2).

Relative Contribution of Each Fiber Type to the Muscle Cross-Sectional Area

Figure 2 shows the relative contribution of each fiber type to the cross-sectional area of the middle block of each muscle. The overall percentage contributed by slow fibers was as follows: quadratus femoris (95%), capsularis (87%),

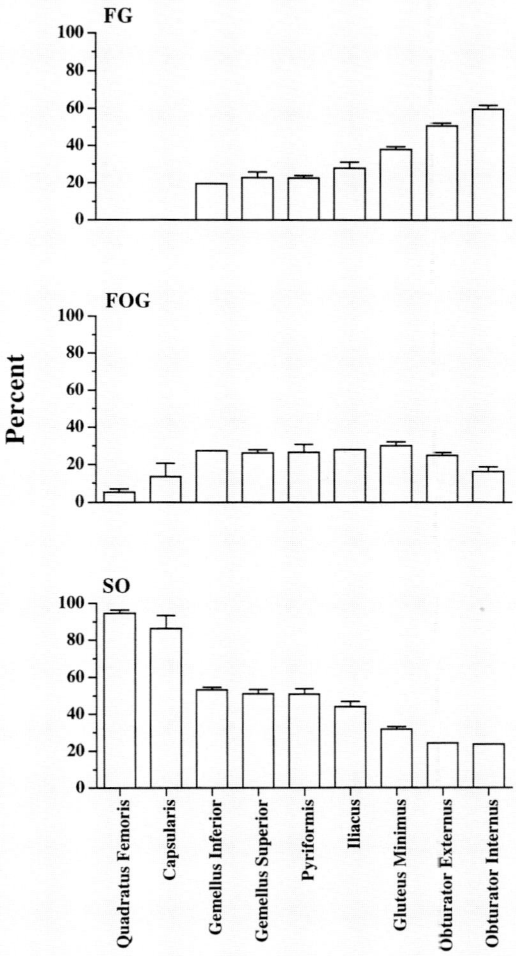

Fig. 2. The relative contribution of each fiber type to the whole muscle cross-sectional area. FG = Fast glycolytic; FOG = fast oxidative glycolytic; SO = slow oxidative. Bars = SEM.

gemellus inferior (53%), gemellus superior (51%), pyriformis (51%), iliacus (44%), gluteus minimus (32%), obturator externus (24%) and obturator internus (24%), respectively.

Discussion

The cat hip 'cuff' muscles comprise a deep layer of relatively small muscles surrounding the hip capsule. As a group, the muscles are relatively short, have short mean fiber lengths, small fiber length:muscle length ratios and low angles of pinnation [compare with Sacks and Roy, 1982]. Overall, the muscles have a relatively high percentage of slow and/or oxidative fibers relative to the majority of other cat hindlimb muscles [Ariano et al., 1973]. Generally, these data are consistent with their presumed function in hip stabilization and support our initial hypothesis as stated in the 'Introduction'.

The architectural design of the cat hip 'cuff' muscles have not been described previously. En bloc staining for motor-end plates using the techniques described by Buckley and Heaton [1968] showed that in all muscles, except the iliacus, there was a single band of innervation for each grossly identified compartment of each muscle (data not shown). In contrast, the iliacus showed multiple motor-end plate bands, suggesting an in series arrangement of fibers similar to that reported previously for relatively long, strap-like muscles in the cat [Adrian, 1925; English and Weeks, 1987; Loeb et al., 1987]. Muscles having relatively long fibers (whether anatomically a single fiber or a series of shorter fibers arranged in series that contract simultaneously) optimize length excursions (displacement or velocity of contraction), whereas muscles having relatively short fibers optimize force production [Gans and Bock, 1965; Josephson, 1975; Spector et al., 1980; Bodine et al., 1982; Gans, 1982; Sacks and Roy, 1982; Edgerton et al., 1983; Powell et al., 1984; Lieber and Blevins, 1989; Peters, 1989; Roy and Edgerton, 1992; Burkholder et al., 1994]. The length of the fibers in all muscles studied, except the iliacus, ranged from 5 to 16 mm and are generally shorter than those reported for a large number of cat hindlimb muscles [see table 1 in Sacks and Roy, 1982]. Thus, it appears that these 'cuff' muscles are designed to optimize force, a feature which would be consistent with a hip stabilization function.

To our knowledge, this is the first description of the fiber type composition of the hip 'cuff' muscles in the cat. Armstrong et al. [1982] reported the fiber-type composition of several of these muscles in the dog, a species in which the fibers can be classified as types I (SO) and IIA (FOG). The overall percent of type I fibers ranged from 37 to 100%: obturator internus (37%), obturator externus (50%), deep gluteal (gluteus minimus) (52%), gemelli (69%), iliacus (70%) and quadratus femoris (100%). These relative values across muscles are in close agreement with those for the cat (table 2). For example, when the percent fiber type composition for each region is weighted by the area of the cross-section that it represents (fig. 2), the overall relative contribution of SO (presumable type I) fibers ranged from 24 to 95%: obturator internus and externus (24%), gluteus minimus (32%), iliacus (44%), pyriformis (51%), gemellus superior (51%), gemellus inferior (53%), capsularis (87%) and quadratus femoris (95%). Interestingly, these absolute values are remarkably close for the cat and dog if the combined contribution for SO and FOG fibers is used for the cat muscles: obturator internus (40%) and obturator externus (49%), gluteus minimus (62%), iliacus (72%), gemellus superior (77%), gemellus inferior (80%), and quadratus femoris (100%). In the rat, Delp and Duan [1996] reported that the contribution of slow fibers to the cross-sectional area of the quadratus femoris, iliacus and obturator internus was about 93, 15 and 11% slow fibers, respectively. The lower percentage of slow fibers in 2/3 rat muscles studied compared to the cat (present data) or dog [Armstrong et al., 1982] may be related to the greater need for faster fibers in smaller animals because of their faster velocities of contraction during locomotion [Hamalainen and Pette, 1993].

The high percentage of slow and oxidative fibers in the hip 'cuff' muscles is consistent with their location close to the bone and/or joint capsule. A layering arrangement of fast to slow from the superficial to the deep muscles is common among the muscles of the hindlimb. For example, in the triceps surae and the quadriceps muscle groups, the deeply located soleus and vastus intermedius muscles, respectively, have the highest percentage of slow fibers among these synergistic groups [Ariano et al., 1973; Armstrong et al., 1982; Armstrong and Phelps, 1984; Acosta and Roy, 1986; also see table 1 in Roy et al., 1988 for a summary].

This arrangement of fiber types is also common within individual muscles. For example, the majority of muscles studied in a variety of mammalian species have a mixture of fiber types and, generally, have a higher proportion of slow and/or high oxidative fibers in the deeper (closer to the bone) than the more superficial (further from the bone) regions of the muscle [Armstrong et al., 1982; Armstrong and Phelps, 1984; Roy et al., 1986; West et al., 1986]. A superficial to deep increase in the percentage of slow fibers was observed in the gemelli superior and inferior and the obturator externus. In several of these muscles (the obturators externus and internus and gemellus inferior and to some degree the quadratus femoris), however, there was a higher percentage of slow fibers in the rostral compared to the caudal portion of the muscle. Assuming a close relationship between the percentage of slow fibers and recruitment level of a muscle portion (see below), these data suggest that the rostral portions of these muscles are activated more frequently than the caudal portions. Electromyographic recordings will be necessary to confirm this contention.

The absolute and relative fiber type compositions of a muscle or a muscle portion appear to reflect, to some extent, its normal chronic activity pattern, i.e. the electrical activation and/or force pattern encountered by the muscle. In this context, muscles which are strongly recruited during postural and low-level activity, e.g. the soleus muscle of the cat [Walmsley et al., 1978; Whiting et al., 1984], rat [Roy et al., 1991] or guinea pig [Gardiner et al., 1982], have a very high percentage of slow fibers. In contrast, muscles that are recruited only during high activity periods, e.g. the medial gastrocnemius [Walmsley et al., 1978] and lateral gastrocnemius [Alaimo et al., 1984] of the cat and the medial gastrocnemius [Roy et al., 1985] and plantaris [Gardiner et al., 1986] of the rat, or muscles whose functional demands appear to be minimal during almost all activity levels, e.g. the tibialis anterior of the cat [Pierotti et al., 1989] and rat [Roy et al., 1985] are composed predominantly of fast fibers.

In all muscles studied, the fiber type composition for each region was similar for the proximal and middle levels and in many cases for the distal level. The variability in the fiber type composition at the distal level of some regions in some muscles, e.g. the obturator internus, gemellus superior and gluteus minimis, is not surprising since these muscles taper and, therefore, have fewer fibers from proximal to distal. Whether this arrangement is due to the narrowing of specific compartments along the proximodistal axis of these muscles as shown by English and Letbetter [1982a, b] for the cat lateral gastrocnemius and plantaris was not addressed in the present study.

Although the significance of intramuscle fiber type regionalization is not fully understood, some structure-function interrelationships have been examined. For example, English and co-workers [English and Letbetter, 1982a, b; English, 1984] have shown an internal structural suborganization of the cat plantaris and lateral gastrocnemius muscles into 'neuromuscular compartments', each compartment consisting of the muscle fibers innervated by a primary branch of the nerve to the muscle and each having a distinct fiber type composition. Further, based on intramuscular electromyographic recordings, it appears that there are variations in the recruitment pattern of individual compartments in the lateral gastrocnemius during locomotor tasks [English, 1984; although see Dunbar and MacPherson, 1993]. Chanaud et al. [1991a, b] have reported differential regional activation patterns in the cat biceps femoris and semitendinosus muscles that appeared to be related to the fiber type composition of the selected regions [also see Bodine et al., 1982; English et al., 1987]. Similarly, data from our laboratory [Roy et al., 1991] involving the recording of the electromyographic activity of the deep and superficial portions of the rat medial gastrocnemius during standing, treadmill running at various speeds and grades and swimming support these contentions. Fibers located in the deep region of the muscle were active during all movements, whereas activity of the fibers in the superficial region was recorded only at high work intensities. Further, De Ruiter et al. [1996] reported a higher percentage of glycogen depleted IIx and IIb fibers in the distal (primarily deep) compared to the proximal (primarily superficial) compartment of the rat medial gastrocnemius during treadmill running at various speeds. Together, these data indicate that the observed regional distributions of fiber types have functional implications.

The cat hip 'cuff' muscles appear to play a role in the control of locomotion and posture. For example, a kinematic analysis of the hindlimb muscles of the cat during ambulatory movements led Grillner and colleagues [Andersson et al., 1978] to suggest that a 'hip trigger' activated during the extension phase of the step cycle reflexly initiated the reversal in movement of the leg. This 'hip trigger' appears to be position sensitive [Grillner and Rossignol, 1978] and most likely involves sensory feedback from the proprioceptors located in the muscles at the hip joint or in the hip joint itself [Grillner, 1975, 1981]. Based on observations of the number of muscle spindles at the three muscle levels studied, it appears that several of the 'cuff' muscles have high spindle densities [Eldred and Roy, unpubl. obs.]. In one muscle, the capsularis, the spindle density has been found to be one of the highest among all cat limb muscles studied [Eldred et al., 1997; see table 1 in Botterman et al., 1978]. Clearly, the capsularis undergoes lengthening during extension of the hindlimb, and thus could play a sensory role in monitoring of flexion-extension at the hip joint. The high percentage of slow fibers, which is highly correlated with high muscle spindle densities, and the location of the other 'cuff' muscles close to the hip joint suggest that these muscle may also function as 'kinesiological monitors' providing important proprioceptive feedback to the central nervous system.

Acknowledgements

The authors thank a large number of undergraduate students for their help in collecting these data. This work was supported, in part, by NIH Grant NS16333.

References

Adrian, E.D. (1925) The spread of activity in the tenuissimus muscle of the cat and in other complex muscles. J Physiol (Lond) *60:* 301–315.

Alaimo, M.A., J.L. Smith, R.R. Roy, V.R. Edgerton (1984) EMG activity of slow and fast ankle extensors following spinal cord transection. J Appl Physiol *56:* 1608–1613.

Andersson, O.S., S. Grillner, M. Lindquist, M. Zomlefer (1978) Peripheral control of the spinal pattern generators for locomotion in cat. Brain Res *150:* 503–507.

Ariano, M.A., R.B. Armstrong, V.R. Edgerton (1973) Hindlimb muscle fiber populations of five mammals. J Histochem Cytochem *21:* 51–55.

Armstrong, R.B., R.O. Phelps (1984) Muscle fiber type composition of the rat hindlimb. Am J Anat *171:* 259–272.

Armstrong, R.B., C.W. Saubert IV, H.J. Seeherman, C.R. Taylor (1982) Distribution of fiber types in locomotory muscles of dogs. Am J Anat *163:* 87–98.

Barka, T., P. Anderson (1963) Histochemistry: Theory, Practice and Bibliography. New York, Harper & Row.

Bodine, S.C., R.R. Roy, D.A. Meadows, R.F. Zernicke, R.D. Sacks, M. Fournier, V.R. Edgerton (1982) Architectural, histochemical, and contractile characteristics of a unique biarticular muscle: The cat semitendinosus. J Neurophysiol *48:* 192–201.

Botterman, B.R., M.D. Binder, D.G. Stuart (1978) Functional anatomy of the association between motor units and muscle receptors. Am Zool *18:* 135–152.

Buckley, G.A., J. Heaton (1968) A quantitative study of cholinesterase in myoneural junctions from rat and guinea-pig extraocular muscles. J Physiol (Lond) *199:* 743–749.

Burkholder, T.J., B. Fingado, S. Baron, R.L. Lieber (1994) Relationship between muscle fiber types and sizes and muscle architectural properties in the mouse hindlimb. J Morphol *221:* 177–190.

Chanaud, C.M., J.M. MacPherson (1991a) Functionally complex muscles of the cat. III. Differential activation within biceps femoris during postural perturbations. Exp Brain Res *85:* 271–280.

Chanaud, C.M., C.A. Pratt, G.E. Loeb (1991b) Functionally complex muscles of the cat hindlimb. V. The roles of histochemical fiber-type regionalization and mechanical heterogeneity in differential muscle activation. Exp Brain Res *85:* 300–313.

Close, R.I. (1972) Dynamic properties of mammalian skeletal muscles. Physiol Rev *52:* 129–197.

Crouch, J.E. (1969) Text-Atlas of Cat Anatomy. Philadelphia, Lea & Febiger.

Delp, M.D., C. Duan (1996) Composition and size of type I, IIA, IID/X, and IIB fibers and citrate synthase activity of rat muscle. J Appl Physiol *80:* 261–270.

De Ruiter, C.J., P.E.M.H. Habets, A. De Haan, A.J. Sargeant (1996) In vivo IIX and IIB fiber recruitment in gastrocnemius muscle of the rat is compartment related. J Appl Physiol *81:* 933–942.

Dunbar, D.C., J.M. MacPherson (1993) Activity of neuromuscular compartments in lateral gastrocnemius evoked by postural corrections during stance. J Neurophysiol *70:* 2337–2349.

Edgerton, V.R., R.R. Roy, R.J. Gregor, C.L. Hager, T. Wickiewicz (1983) Muscle fiber activation and recruitment; in Knuttgen H, Vogel J, Poortmans J (eds): Biochemistry of Exercise, International Series of Sport Sciences, vol 13. Illinois, Human Kinetics, pp 31–49.

Eldred, E., M. Ounjian, R.R. Roy, V.R. Edgerton (1993) Tapering of the intrafascicular endings of muscle fibers and its implications to relay of force. Anat Rec *236:* 390–398.

Eldred, E., L. Yung, R.R. Roy (1997) Spindle representation relative to distribution of muscle fiber types in the cat capsularis muscle. Acta Anat *159:* 114–126.

English, A.W. (1984) An electromyographic analysis of compartments in cat lateral gastrocnemius during unrestrained locomotion. J Neurophysiol *52:* 114–125.

English, A.W., W.D. Letbetter (1982a) A histochemical analysis of identified compartments of cat lateral gastrocnemius muscle. Anat Rec *204:* 123–130.

English, A.W., W.D. Letbetter (1982b) Anatomy and innervation patterns of cat lateral gastrocnemius and plantaris muscles. Am J Anat *164:* 67–77.

English, A.W., O.I. Weeks (1987) An anatomical and functional analysis of cat biceps femoris and semitendinosus muscles. J Morphol *191:* 161–175.

Gans, C. (1982) Fiber architecture and muscle function. Exerc Sport Sci Rev *10:* 160–207.

Gans, C., W.J. Bock (1965) The functional significance of muscle architecture: A theoretical analysis. Ergeb Anat Entwicklungsgesch *38:* 115–142.

Gans, C., F. de Vree (1987) Functional bases of fiber length and angulation in muscle. J Morphol *192:* 63–85.

Gardiner, K.R., P.F. Gardiner, V.R. Edgerton (1982) Guinea pig soleus and gastrocnemius electromyograms at varying speeds, grades, and loads. J Appl Physiol *52:* 451–457.

Gardiner, P., R. Michel, C. Bowman, E. Noble (1986) Increased EMG of rat plantaris during locomotion following surgical removal of its synergists. Brain Res *380:* 114–121.

Gordon, A.M., A.F. Huxley, F.J. Julian (1966) The variation in isometric tension with sarcomere length in vertebrate muscle fibers. J Physiol (Lond) *184:* 170–192.

Grillner, S. (1975) Locomotion in vertebrates: Central mechanisms and reflex interaction. Physiol Rev *55:* 247–304.

Grillner, S. (1981) Control of locomotion in bipeds, tetrapods, and fish; in Brookhart JM, VB Mountcastle (eds): Handbook of Physiology, The Nervous System, Motor Control, vol 2, part 1, sect 1. Bethesda, American Physiological Society, pp 1179–1236.

Grillner, S., S. Rossignol (1978) On the initiation of the swing phase of locomotion in chronic spinal cats. Brain Res *146:* 269–277.

Hamalaineen, N., D. Pette (1993) The histochemical profiles of fast fiber types IIB, IID, and IIA in skeletal muscle of mouse, rat, and rabbit. J Histochem Cytochem *41:* 733–743.

Josephson, R.K. (1975) Extensive and intensive factors determining the performance of striated muscle. J Exp Zool *194:* 135–154.

Lieber, R.L., F.T. Blevins (1989) Skeletal muscle architecture of the rabbit hindlimb: Functional implications of muscle design. J Morphol *199:* 93–101.

Loeb, G.E., C.A. Pratt, C.M. Chanaud, F.J. Richmond (1987) Distribution and innervation of short, interdigitated muscle fibers in parallel-fibered muscles of the cat hindlimb. J Morphol *191:* 1–15.

Mendez, J., A. Keys (1960) Density and composition of mammalian muscle. Metabolism *9:* 184–188.

Muhl, Z.F. (1982) Active length-tension relation and the effect of muscle pinnation on fiber lengthening. J Morphol *173:* 285–292.

Nwoye, L., W.F. Mommaerts, D.R. Simpson, K. Seraydarian, M. Marusich (1982) Evidence for a direct action of thyroid hormone in specifying muscle properties. Am J Physiol *242:* R401–R408.

Otten, E. (1988) Concepts and models of functional architecture in skeletal muscle. Exerc Sport Sci Rev *16:* 89–137.

Ounjian, M., R.R. Roy, E. Eldred, A. Garfinkel, J.R. Payne, A. Armstrong, A.W. Toga, V.R. Edgerton (1991) Physiological and developmental implications of motor unit anatomy. J Neurobiol *22:* 547–559.

Pierotti, D.J., R.R. Roy, R.J. Gregor, V.R. Edgerton (1989) Electromyographic activity of cat hindlimb flexors and extensors during locomotion at varying speeds and inclines. Brain Res *481:* 57–66.

Peter, J.B., R.J. Barnard, V.R. Edgerton, C.A. Gillespie, K.E. Stempel (1972) Metabolic profiles of three fiber types of skeletal muscle in guinea pigs and rabbits. Biochemistry *11:* 2627–2633.

Peters, S.E. (1989) Structure and function in vertebrate skeletal muscle. Am Zool *29:* 221–234.

Powell, P.L., R.R. Roy, P. Kanim, M.A. Bello, V.R. Edgerton (1984) Predictability of skeletal muscle tension from architectural determinations in guinea pig hindlimbs. J Appl Physiol *57:* 1715–1721.

Richmond, F.J.R., J.B. Armstrong (1988) Fiber architecture and histochemistry in the cat neck muscle, biventer cervicis. J Neurophysiol *60:* 46–59.

Roy, R.R., L. Acosta, Jr. (1986) Fiber type and fiber size changes in selected thigh muscles six months after low thoracic spinal cord transection in adult cats: Exercise effects. Exp Neurol *92:* 675–685.

Roy, R.R., K.M. Baldwin, V.R. Edgerton (1991a) The plasticity of skeletal muscle: Effects of neuromuscular activity. Exerc Sports Sci Rev *19:* 269–312.

Roy, R.R., K.M. Baldwin, T.P. Martin, S.P. Chimarusti, V.R. Edgerton (1985) Biochemical and physiological changes in overloaded rat fast and slow ankle extensors. J Appl Physiol *59:* 639–646.

Roy, R.R., M.D. DiGiuro, V.R. Edgerton (1983) Architectural design of adult cat hip musculature. Soc Neurosci Abstr *9:* 359.

Roy, R.R., V.R. Edgerton (1992) Skeletal muscle architecture and performance; in Komi PV (ed): Strength and Power in Sport: Encyclopedia of Sports Medicine, vol 3. Oxford, Blackwell Scientific, pp 115–129.

Roy, R.R., D.L. Hutchison, D.J. Pierotti, J.A. Hodgson, V.R. Edgerton (1991b) EMG patterns of rat ankle extensors and flexors during treadmill locomotion and swimming. J Appl Physiol *70:* 2522–2529.

Sacks, R.D., R.R. Roy (1982) Architecture of the hind limb muscles of cats: Functional significance. J Morphol *173:* 185–195.

Spector, S.A., P.F. Gardiner, R.F. Zernicke, R.R. Roy, V.R. Edgerton (1980) Muscle architecture and the force velocity characteristics of the cat soleus and medial gastrocnemius: Implications for motor control. J Neurophysiol *44:* 951–960.

Trotter, J.A. (1991) Dynamic shape of tapered skeletal muscle fibers. J Morphol *207:* 211–223.

Trotter, J.A. (1993) Functional morphology of force transmission in skeletal muscle. Acta Anat *146:* 205–222.

Walmsley, B., J.A. Hodgson, R.E. Burke (1978) Forces produced by medial gastrocnemius and soleus muscles during locomotion in freely moving cats. J Neurophysiol *41:* 1203–1215.

Wattenberg, L.W., J.L. Leong (1960) Effects of coenzyme Q10 and menadione on succinate dehydrogenase activities as measured by tetrazolium salt reduction. J Histochem Cytochem *8:* 296–303.

West, S.P., R.R. Roy, V.R. Edgerton (1986) Fiber type and fiber size of cat ankle, knee, and hip extensors and flexors following low thoracic spinal cord transection at an early age. Exp Neurol *91:* 174–182.

Whiting, W.C., R.J. Gregor, R.R. Roy, V.R. Edgerton (1984) A technique for estimating mechanical work of individual muscles in the cat during treadmill locomotion. J Biomech *17:* 685–694.

Wickiewicz, T.L., R.R. Roy, P.L. Powell, J.J. Perrine, V.R. Edgerton (1984) Muscle architecture and force-velocity relationships in humans. J Appl Physiol *57:* 435–443.

Acta Anatomica

Acta Anat 1997;159:147–155

G. S. Chleboun[a]
T. J. Patel[b]
R. L. Lieber[b]

[a] Ohio University School of Physical Therapy, Athens, Ohio, and
[b] Departments of Orthopaedics and Bioengineering, Biomedical Sciences Graduate Group, University of California and Veterans Administration Medical Centers, San Diego, Calif., USA

Skeletal Muscle Architecture and Fiber-Type Distribution with the Multiple Bellies of the Mouse Extensor digitorum longus Muscle

Key Words
Extensor digitorum longus
Multitendon
Architecture
Fiber type
Compartmentalization

Abstract
The purpose of this study was to describe the extent to which architectural and fiber-type characteristics of the four bellies of the mouse extensor digitorum longus (EDL) suggest specialization of the digits, and to mathematically model the functional effects of the structural properties. Six mice were perfused in situ with glutaraldehyde while the lower limb was positioned approximately in the neutral position. After perfusion, lower limbs were removed and placed in glutaraldehyde until the EDL was dissected from the limb and separated into individual muscle bellies corresponding to each digit for architectural determination. The results showed that the muscle belly of digit 5 tended to be different from the muscle bellies of digits 2–4 for many architectural characteristics. Muscle mass, physiological cross-sectional area, muscle length, and fiber length were all significantly greater in digit 5. Proximal tendon length was also significantly longer in digit 5, and distal tendon length, as well as total tendon length, were significantly shorter in digit 5. Sarcomere length was shortest at the proximal end of the muscle and longest, 60–80%, toward the distal end. Fiber type distribution was about 60% FOG, 39% FG with only 1% SO fibers in all muscle bellies. Muscle-tendon modeling illustrated that peak force and maximal shortening velocity were greatest in digit 5. Inclusion of the tendon in the model resulted in a 10% shift of the force-length curve to longer lengths. Assuming muscle structure is matched to function, we speculate that digit 5 of the mouse EDL bears higher loads over a greater excursion during locomotion compared to the remaining digits.

Introduction

The musculoskeletal system presents numerous examples of structural specialization required to accomplish a particular task. For example, specific skeletal muscle fiber orientations within whole muscles (i.e. muscle architecture) uniquely suit certain muscles to perform tasks where high excursions, high forces or both are required [Gans, 1982; Sacks and Roy, 1982; Wickiewicz et al., 1983; Powell et al., 1984]. In the same way, depending on the pattern of use within a particular muscle, fiber-type distributions are appropriate for certain muscles to perform tasks that require

KARGER

Fax +41 61 306 12 34
E-Mail karger@karger.ch
www.karger.com

This article is also accessible online at:
http://BioMedNet.com/karger

Richard L. Lieber, PhD
Department of Orthopaedics (9151)
U.C. San Diego School of Medicine and V.A. Medical Center
3350 La Jolla Village Drive, San Diego, CA 92161 (USA)
Tel. +1 619 552-8585, ext. 7016, Fax +1 619 552-4381, E-Mail rlieber@ucsd.edu

high endurance or low endurance. Thus, it is not surprising to find variations in fiber-type distributions across muscles where different quantities or qualities of activity are required. Finally, with regard to the muscle-tendon unit itself, tendon mechanical properties and dimensions vary widely among muscles. For example, the digital flexors have extremely long tendons that are very stiff suiting them for precise positional control of the fingers [Lundberg et al., 1997] while at the other extreme, the Achilles tendon of the lower limb is a relatively compliant and extremely long tendon, well-suited for energy storage. At least in part, this increases the efficiency of locomotion in cats, wallabies, and humans [Cavagna and Kaneko, 1977; Morgan et al., 1978; Hoffer et al., 1989]. Taken together, variations in muscle architecture, fiber-type distribution and tendon geometry present the musculoskeletal system with the ability to create customized muscle-tendon units suited to perform a particular task.

It has been demonstrated in a number of human [Wickiewicz et al., 1983; Lieber et al., 1992] and animal systems [Sacks and Roy, 1982; Burkholder et al., 1994] that muscle-tendon unit specialization occurs between functional groups. For example, the plantar flexors of the lower limb are typically muscles composed of relatively short fibers producing relatively high forces in contrast to the dorsiflexors that produce relatively high excursions and lower forces. These designs seem consistent with each muscle groups' intended function. Similarly, the primary knee extensors, the quadriceps femoris, are typically short-fibered muscles that generate relatively high forces in contradistinction to the primary knee flexors, the hamstrings, that have higher excursions and produce relatively lower forces. Extending this concept further, numerous muscles have multiple bellies or multiple compartments that may indicate that they have truly separate functions or may simply represent geometrical constraints imposed upon the system. From an architectural point of view, specialization within the human digital flexors and extensors was demonstrated by the classic studies of Brand et al. [1981] and later amplified by Lieber et al. [1992]. However, to date, the relationship between skeletal muscle architecture, fiber-type distribution, and tendon dimensions has not been studied in a multibellied muscle. Thus, the purpose of this study was to measure the architecture, fiber-type distribution and tendinous properties of the mouse extensor digitorum longus (EDL) muscle to understand the relative specialization of the multiple bellies within the muscle.

Methods

Six Swiss Webster adult female mice (range 20–22 g ByW) were anesthetized with a cocktail (xylazine 5 mg/kg, acepromazine 1 mg/kg, and ketamine 50 mg/kg) intraperitoneally. The animals were positioned supine for in situ perfusion fixation with the knees in 90° flexion and ankles in 125° plantar flexion approximating neutral position while under anesthesia. The knees, ankles, and feet were held in place with molding clay in a paraffin splint. Perfusion fixation in situ was performed as described previously [Mathieu-Costello, 1987]. Briefly, the heart was exposed and 0.2 ml of papaverine hydrochloride was injected into the left ventricle as a vasodilator. A catheter was inserted into the left ventricle and the right atrium was clipped to allow outflow. The vascular system was flushed with heparinized saline (2 ml/l) for 1.5 min at a constant pressure of 120 mm Hg. Immediately following the perfusion with saline, the vascular system was perfused with a 6.25% solution of glutaraldehyde in 0.1 *M* sodium cacodylate buffer for 10 min at 120 mm Hg to fix the tissues.

The lower limbs were removed and refrigerated in glutaraldehyde solution for 72 h. The right lower limbs were then rinsed and placed in buffer until dissection of the EDL. Prior to the dissection, radiographs of the limbs were taken. Knee and ankle joint angles were measured using a standard plastic goniometer aligning the arms of the goniometer with the long axis of the femur, tibia, and metatarsals. Using a dissection microscope, the EDL from the proximal attachment at the femur to the metatarsal-phalangeal joint was removed. Muscle mass and muscle length of the whole EDL were measured prior to separation of the muscle bellies. The muscles were blotted dry and muscle mass was measured with an analytical balance. Muscle length, the distance from most proximal to most distal fibers, was measured with calipers under the dissection microscope. The muscle bellies of the four digits (digits 2–5) were separated by blunt dissection and muscle belly orientation was noted. Relatively small amounts of connective tissue joined the bellies permitting clean dissection of individual bellies and measurement of their architectural properties. Muscle mass, muscle length, proximal tendon length, and distal tendon length were measured on each of the digits. Pennation angle of each muscle belly was measured by placing the muscle on a piece of clear plastic with angle markings etched into the surface with a scalpel blade. Single fibers or small bundles of fibers (2–10 fibers) were teased apart from the lateral, middle, and medial portions of each muscle belly. In single fibers or fiber bundles of 2–4 fibers, fiber length was considered the length of the bundle as measured with calipers under appropriate magnification. When there were as many as 5–10 fibers per bundle, care was taken to measure the length of the fibers at the periphery of the bundle since the fibers entered the tendon in a slightly staggered array. Muscle length and fiber length were corrected to a sarcomere length of 2.5 μm to compensate for differences in joint angle during fixation. The fiber bundles were then mounted onto a microscope slide being careful to maintain the bundles in a straight line for measurement of sarcomere length. Physiological cross-sectional area (PCSA) for each digit was calculated using the equation [Sacks and Roy, 1982]:

$$\mathrm{PCSA}\ (\mathrm{mm}^2) = (M(\mathrm{g}) \cdot \cos\theta)/(\rho(\mathrm{g/mm}^3) \cdot L_f\ (\mathrm{mm})),$$

where M is muscle mass, θ is pennation angle, ρ is the density of muscle (0.001056 g/mm^3 [Mendez and Keys, 1960], and L_f is the fiber length).

Sarcomere length was measured at 0.25-mm increments along the fiber bundles using laser diffraction [Lieber et al., 1984]. Based on the distance between the zeroth and first order diffraction lines, sarcomere length was calculated from the equation: $n\lambda = d\sin\theta$, where n = diffraction order, λ = laser wavelength (0.632 μm), d = sarcomere length, θ = diffraction angle relative to the undiffracted beam. The fiber bundle on the slide was aligned with a custom stage attached to a micrometer that allowed precise translation of the laser beam along the fiber axis. Sarcomere length measurements were always performed proximal to distal beginning at the point of the first observable diffraction pattern as close to the proximal tendon as possible (usually about 0.5 mm, half the width of the laser beam).

Fiber-type determination was performed using EDL muscles from three separate Swiss Webster mice (range 24.7–25.0 g). The mice were anesthetized as described above and the EDL muscles from both hind limbs were immediately removed and placed in chilled Ringer's solution. The muscle bellies of each of the four digits were carefully separated from distal to proximal and placed in parallel between two layers of porcine gel to minimize freezing artifact. The gel-muscle sandwich was then frozen in isopentane cooled by liquid nitrogen (–159 °C), and stored at –80 °C. Serial sections (10 μm) were stained with myofibrillar ATPase in alkaline preincubation (pH 10.4), succinate dehydrogenase (SDH) and α-glycerophosphate dehydrogenase. Myofibrillar ATPase was used to distinguish between fast and slow fiber types, and an acidic preincubation (pH 4.25) confirmed the reversal of staining pattern under alkaline and acidic preincubation. SDH was used to distinguish between oxidative and non-oxidative fiber types and α-glycerophosphate dehydrogenase [Pearse, 1961] to distinguish between glycolytic and non-glycolytic fibers. The fibers were then classified as either slow oxidative (SO), fast oxidative glycolytic (FOG), or fast glycolytic (FG) [Peter et al., 1972].

Modeling of Muscle-Tendon Properties

Estimates of maximal isometric tension and maximum shortening velocity for each whole EDL and each digit's muscle belly were obtained using a mathematical model similar to that of Zuurbier and Huijing [1992] and described in detail elsewhere [Burkholder et al., 1994]. Muscle mass, fiber length, muscle length, pennation angle, and fast fiber type percentage were used to model the contractile properties of the muscles. The sarcomere force-length and force-velocity properties were taken from values reported in the literature [Katz, 1939; Gordon et al., 1966; Close, 1972]. The plateau for the force-length relationship was set to 2.47–2.81 μm based on data from mammalian muscle, and the value used for specific tension of the muscle was 22.5 $N\text{-}cm^{-2}$ [Close, 1972].

Statistical Analysis

Differences between digits for all architectural and morphological variables were analyzed using one-way analysis of variance (ANOVA), and multiple comparisons were made using Tukey's Honestly Significant Difference (HSD) test ($\alpha = 0.05$). Sarcomere length along the fiber was analyzed using repeated measures ANOVA, and multiple comparisons (50 comparisons) between positions along the fiber were analyzed using Student's t tests with the Bonferonni correction for α ($\alpha = 0.05/50 = 0.001$). For statistical analysis of sarcomere length along the fiber, a subset of 10 fibers in which sarcomere length measurements could be obtained along the entire fiber length was used. This was necessary because, in many cases, a sarcomere length measurement was not possible along the complete length of the fiber because of the lack of striation pattern or imperfection in the mounting of the fibers. The total of 10 fibers used represented fibers from each of the six mice and each of the muscle bellies of the four digits. All data are presented as the mean ± SD.

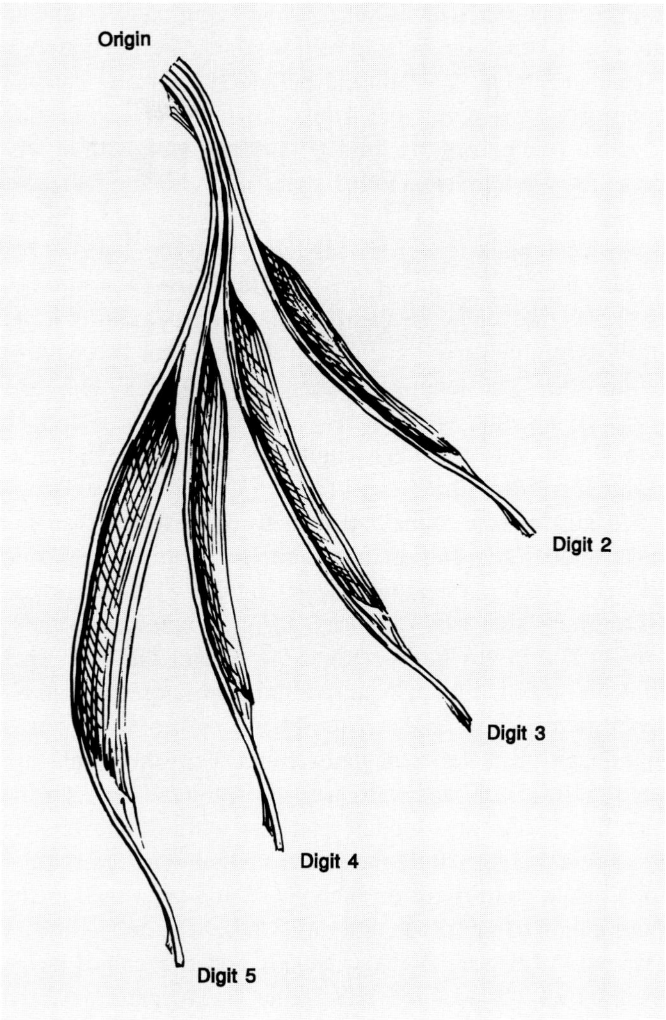

Fig. 1. Diagram of the mouse EDL muscle showing the four distinct muscle bellies and the staggered proximal attachment (complete tendons are not shown). Muscle belly length from digits 2–5 are scaled to the average lengths of 7.0, 7.3, 7.7, and 9.0 mm, respectively.

Results

During fixation the knee and ankle joints were held in a neutral position approximating the position while the animal was under anesthesia. Average angles at the knee and ankle joints were 91.2 ± 3.3° and 124.7 ± 7.9°, respectively. The metatarsophalangeal joints, although slightly flexed in the anesthetized animal, were more easily positioned at a consistent position of 180°. Measurements of muscle mass, muscle length, and calculation of PCSA were made on the whole EDL prior to separating the four digits of the EDL. Whole muscle mass was 7.1 ± 1.4 mg, muscle length was 12.1 ± 0.6 mm, and PCSA was 1.3 ± 0.2 mm^2.

The four digits of the EDL have anatomically distinct and easily separable muscle bellies arising from a common proximal tendon. The muscle bellies are narrow from an anterior view, much wider from a lateral view, and nestled together into an orderly fit. They are arranged with digit 2 being most proximal and medial and digit 5 being most distal and lateral (fig. 1). The proximal attachment of each digit to the main tendon is staggered with the distance between successive digits increasing from about 0.4 mm between digits 2 and 3 to about 1.6 mm between digits 4 and 5. Distally, the tendons attach to the extensor mechanism of the toes. The tendon for digits 2–4 attach only to the respective digits, whereas the tendon for digit 5 has an aponeurosis that extends to digits 2–4 in addition to the attachment to digit 5.

Architecturally, digit 5 tended to be different than the other three digits. Digit 5 had significantly greater mass and PCSA ($p<0.05$) than digits 2–4 (fig. 2). The differences were impressive with the mass of digit 5 being about 40% greater and PCSA being about 31% greater than the other digits. The fifth digit also had a significantly greater muscle length than digits 2–4, and the fiber length of digit 5 was significantly greater that digits 2 and 3 ($p<0.05$), but not digit 4 (fig. 3). Fiber length was not different between the lateral, middle, and medial portions within each muscle ($p>0.05$), therefore only the data from the middle portion of the muscle were used for analysis. When expressed as the fiber length to muscle length ratio, there was no significant difference between digits ($p>0.05$) and the values range from 0.69 ± 0.04 to 0.74 ± 0.04 (table 1). The pattern for the uniqueness of digit 5 was also apparent when comparing the tendon length as the proximal, distal, and total tendon lengths of digit 5 were significantly different ($p<0.05$) than the other digits (fig. 4). The proximal tendon length increased and distal tendon length decreased from digit 2 to digit 5, providing the staggered array of the digits mentioned above. Proximal tendon length of both digits 4 and 5 were significantly longer than digits 2 and 3, and digit 5 was significantly longer than digit 4. The differences between digits were the same for the distal tendon length as those seen in the proximal tendon, but the tendons were shorter in digits 4 and 5. The total tendon length of digit 5 was significantly less than digits 2 and 3, and digit 4 was significantly less than digit 2. The ratio between total tendon length and fiber length decreased from digit 2 to digit 5, which follows from the shorter tendon and longer fiber length in digit 5. Although not significant statistically, the data suggest a trend where digit 5 pennation angle was larger and the angles became progressively smaller from digit 5 to digit 2 (table 1).

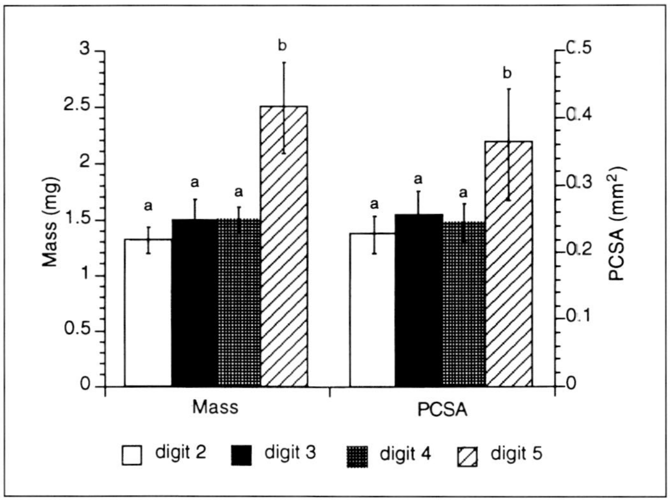

2

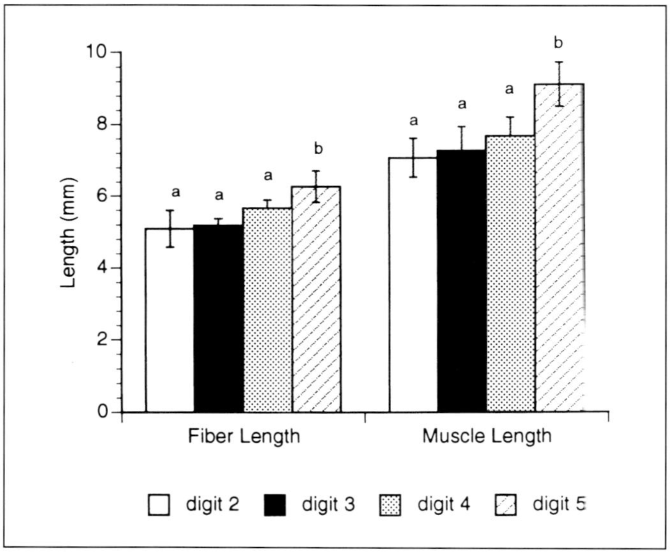

3

Fig. 2. Mass and PCSA of digits 2–5 (±SD). Significant differences between digits are designated by different letters ($p<0.05$).

Fig. 3. Fiber length and muscle length of digits 2–5 (±SD). Fiber length represents the fibers from the middle portion of each of the muscle bellies. Significant differences between digits are designated by different letters ($p<0.05$).

There were no significant differences between digits with respect to fiber-type distribution ($p>0.05$). Each digit contained predominately fast fibers based on ATPase staining (98.4 ± 2.1 to 99.6 ± 0.9% fast contracting fibers, 1.6 ± 2.1 to 0.4 ± 0.9% slow contracting fibers). Oxidative capacity of each digit was about 60% FOG, 39% FG, and 1% SO with no difference between digits (table 1).

Sarcomere length measurements were made along the length of the fiber or fiber bundle. For comparisons be-

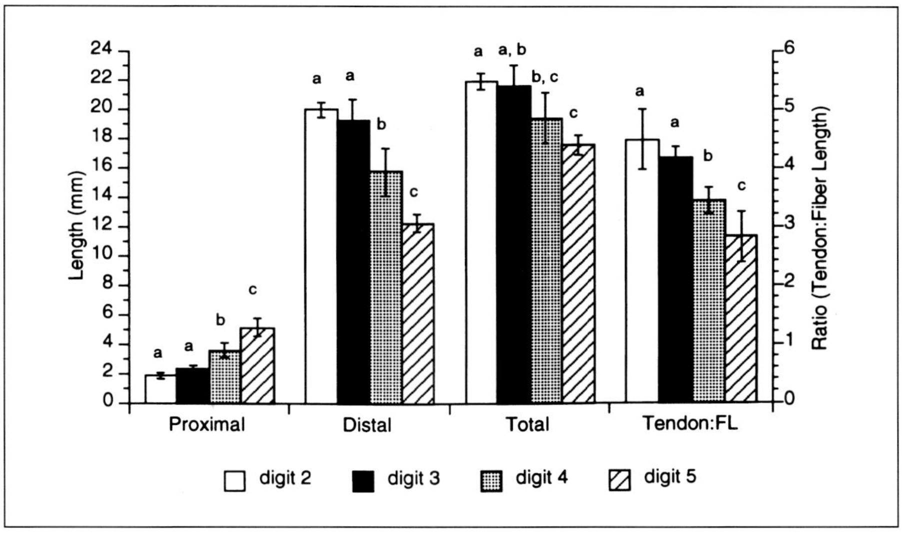

Fig. 4. Tendon length and tendon length:fiber length ratio for digits 2–5 (±SD). Significant differences between digits are designated by different letters ($p<0.05$).

Table 1. Architectural properties, fiber type distribution, and estimated contractile properties from each muscle belly of each mouse EDL digit

Parameter	Digit 2	Digit 3	Digit 4	Digit 5
Pennation angle, °	8.6 ±3.0	9.3 ±1.0	10.6 ±2.4	11.3 ±2.1
L_f:L_m ratio	0.72±0.03	0.72±0.05	0.74±0.04	0.69±0.04
Sarcomere length, μm	2.58±0.09	2.64±0.13	2.63±0.08	2.55±0.12
% FOG fibers	66.1 ±5.3	58.0 ±10.0	58.5 ±10.9	58.7 ±9.5
% FG fibers	33.5 ±5.0	40.4 ±11.7	40.3 ±10.4	39.7 ±9.1
Calculated P_o, g	6.15±1.17[a]	6.88±0.97[a]	6.31±0.86[a]	9.11±1.69[b]
Calculated V_{max}, mm/s	34.62±4.44[a]	36.61±1.32[a]	40.09±1.68[a]	44.36±3.08[b]

Different letters signify significant differences between digits ($p<0.05$).

L_f:L_m=Fiber length:muscle length ratio; FOG=fast oxidative glycolytic; FG=fast glycolytic; P_o=maximum tetanic tension; V_{max}=maximum contraction velocity.

tween lateral, middle, and medial fiber bundles and between digits, the mean sarcomere length measurement for each fiber or fiber bundle was used. There was no statistical difference in sarcomere length between the fiber bundle locations ($p>0.05$), hence only the sarcomere length from the middle portion was used to compare between digits. There was no significant difference between digits ($p>0.05$) with sarcomere lengths ranging from 2.55±0.12 to 2.64±0.13 μm (table 1). Sarcomere length variability along the fiber was evaluated using the subset of 10 fibers and fiber bundles in which sarcomere lengths were obtained for the complete length of the fiber. Since fibers were of varying lengths, relative positions along the fibers were used. The sarcomere length was about 10% longer ($p<0.05$) at the section to the fiber representing 60–80% of the fiber length from the proximal end than the most proximal portion of the fiber (fig. 5).

Mathematical modeling based on gross architectural properties of the whole EDL muscle revealed an average estimated peak tension of 29.2±6.5 g and maximal shortening velocity of 40.4±1.7 mm/s for the whole EDL muscle. Data for estimates of peak tension and maximal shortening velocity for each of the digits are presented in table 1. As expected, digit 5 had the largest estimated peak tension and shortening velocity compared to the other three digits reflecting its greater mass and fiber length, respectively. Estimated force-length (fig. 6A) and force-velocity curves (fig. 6B) clearly reveal the functional dominance of digit 5 over the other digits. When the effect of tendon properties are considered on the force-length characteristics of the different digits, a shift in the force-length curve to the right was seen (fig. 7).

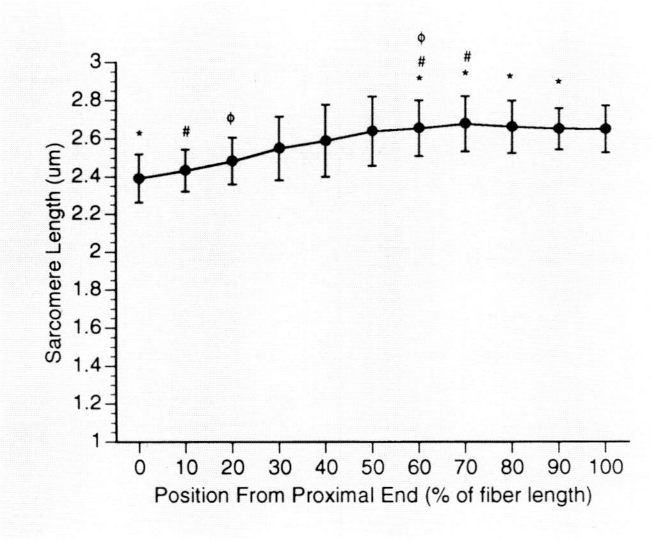

5

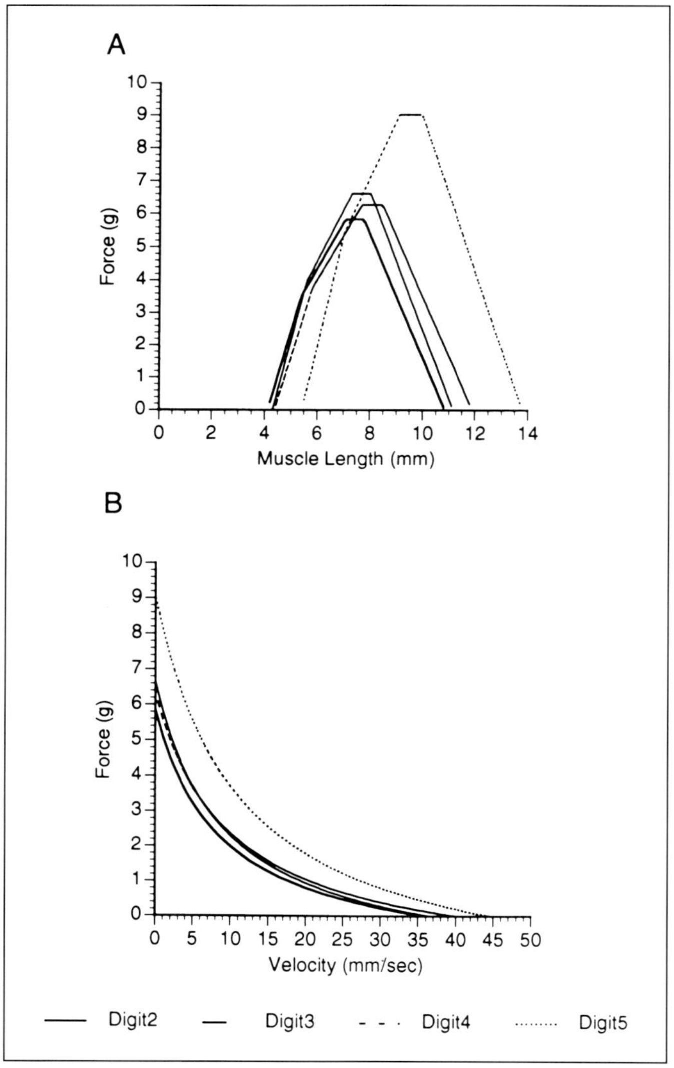

6

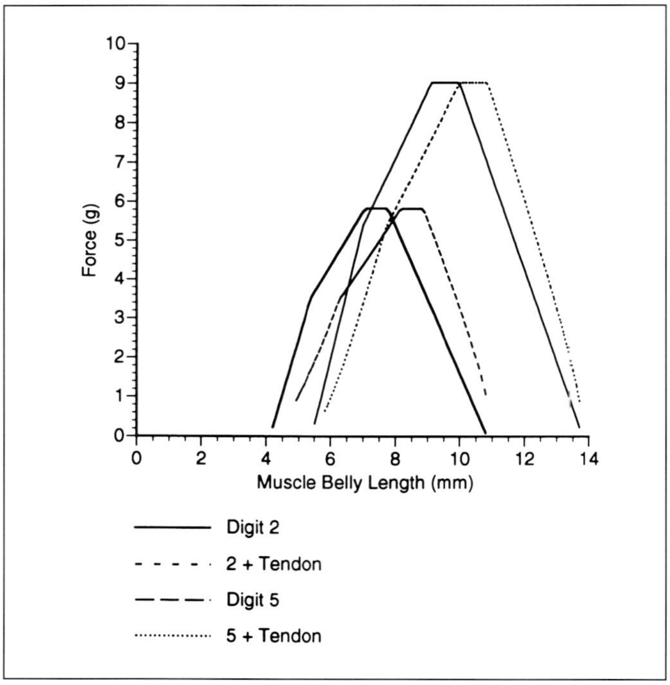

7

Fig. 5. Sarcomere length along the fiber at 10% intervals along the fiber (±SD). Significant differences between positions are designated by similar symbols ($p<0.05$).

Fig. 6. Estimated force-length (**A**) and force-velocity (**B**) relationships for each of the four digits.

Fig. 7. The effect of tendon compliance on the force-length property of digit 2 and digit 5. Note that the shift to the right is slightly greater in digit 2 than digit 5.

Discussion

The anatomical orientation of the mouse EDL represents a unique design in which there are four distinct muscle bellies arising from the shared proximal tendon and four separate distal tendons of insertion. The only overlapping structure is the distal end of digit 5 tendon where it has a broad aponeurosis attaching at the second through fourth metatarsophalangeal joints, but this aponeurosis has no connection to the tendons of digits 2–4. The staggering of the four bellies reflects the differences in proximal and distal tendon lengths that minimizes the width of the whole muscle allowing it to be packed between the tibialis anterior and the fibula.

The architectural characteristics of the EDL, as a whole, are similar to those reported previously [Burkholder et al., 1994]. In addition, fiber length for digit 5 in the present study (6.3±0.4 mm) compares favorably with the fiber length presented in the previous report (6.2±2.2 mm). Fiber length to muscle length ratio is similar between the two studies for the whole EDL (50%); however, for individual digits the ratio is much higher (72%) reflecting the shorter muscle length of the individual digits. These data demonstrate the importance of reducing the whole muscle preparation to individual bellies in establishing actual fiber lengths. Comparing to the extensor hallucis longus (EHL) and the tibialis anterior (TA), the other ankle dorsiflexors, the fiber length to muscle length ratio for individual digits fell between the values for these muscles (81 and 61%, respectively). In fact, digits 2–4 tend to be very architecturally similar to the EHL.

The results clearly show that digit 5 is architecturally distinct compared to digits 2–4. The larger PCSA and longer fiber length of digit 5 suggests this digit has the potential to develop greater force over a greater excursion and have a greater shortening velocity. Estimates of maximal isometric tension and maximal shortening velocity from muscle modeling reflect these architectural differences. Although there are no published reports of physiological measurements of maximal tetanic tension and shortening velocity of the individual digits of the mouse EDL, pilot data from our laboratory suggest that the modeling estimates are a reasonable approximation of the actual physiological properties of the individual digits. Comparing the estimated maximal tension and shortening velocity for the whole EDL to known values is possible, however. Maximal isometric tension has been reported to be about 34–42 g at 25–35 °C which is somewhat higher than our estimate [Luff, 1981; Brooks and Faulkner, 1988; Warren et al., 1994]. Maximum shortening velocity was between 5.9 and 9.8 fiber lengths/s which brackets our estimate of 7.1 fiber lengths/s [Crow and Kushmerick, 1983; Brooks and Faulkner, 1988]. Based on these reports, the model appears to provide a reasonable approximation of the physiological properties of the EDL muscle. Since there are no differences in fiber type distribution between digits, it is clear that muscle architecture dominates in the determination of functional properties of the muscle.

Tendon length also differed across digits. Digit 5 combines the shortest total tendon length with the greatest potential for force development. Assuming similar material properties between the different tendons, the shorter tendon would have less total compliance and provide greater control of movement. The effect of the different tendon lengths is seen in figure 7 where the shift of the force-length curve is slightly greater in digit 2 (longer tendon) than in digit 5 (shorter tendon), even though absolute force is greater in digit 5.

The architectural characteristics of other multitendoned muscles in the forelimb have been described for humans, primates, and cats. The human and primate extensor digitorum communis (EDC) muscle bellies are inseparable for the proximal 2/3 of the muscle and split into four bellies only over the distal 1/3 of the muscle, presenting a very different design from the distinct bellies of the mouse EDL [Lieber et al., 1992; Serlin and Schieber, 1993]. In the human lower extremity, the EDL muscle as a whole has a relatively small fiber length to muscle length ratio (0.23), although the human lower extremity digital flexors and extensors also tend to have lower fiber:muscle length ratios than the mouse [Wickiewicz et al., 1983]. This is also seen in the cat where the whole EDL fiber:muscle length ratio is 0.35 [Sacks and Roy, 1982].

The multiple bellies of the EDL can be considered a functional group of muscles within themselves similar to the general functional group of ankle dorsiflexors. As a functional group, they would be expected to have similar architectural and fiber type characteristics [Burkholder et al., 1994]. As noted above, differences were found between the digits, but when compared to other ankle dorsiflexor muscles (TA and EHL), the differences may not be considered as important. In addition, the pattern of innervation of the EDL suggests that the anatomical separation of the bellies may not be the primary determinant of functional compartmentalization. Balice-Gordon and Thompson [1988] identified two branches of the common peroneal nerve that innervate the proximal-medial and distal-lateral portions of the EDL muscle in rats. Each branch innervates about half of the EDL overlapping the muscular boundaries of the digits. Only digit 5 is innervated by just one branch, the distal-lateral branch. Because the muscular compartments do not match the neuromuscular compartments, individual extension of the toes is not possible and may not be necessary because of the pattern of muscle activity during gait. The EDL is active during the swing phase of gait in the rat [Nicolopoulos-Stournaras and Iles, 1984] presumably to assist in dorsiflexion of the ankle and extension of the toes for ground clearance. In this sense, mass extension of the digits and dorsiflexion of the ankle would be preferable over individualized extension of the digits. The longer fiber length of digit 5 may be related to a greater excursion of digit 5 muscle during gait. In the human ankle, plantar flexion at the end of stance phase is accompanied by inversion of the subtalar joint. This position may result in greater lengthening of digit 5 requiring larger excursion to dorsiflex the foot during swing phase. Whether this is the same in the mouse is not known. The larger size of digit 5 permits higher force generation which may balance the foot laterally as the large toe stabilizes it medially.

The longer fiber length in digit 5 may be a design feature that allows for different 'gearing' between the muscle bellies of the EDL. A change in whole EDL muscle length will result in a greater relative change in sarcomere length of digits 2–4 than digit 5, assuming similar moment arms. As the muscle shortens during dorsiflexion in the swing phase of gait, digits 2–4 may reach the plateau of the length-tension curve prior to when digit 5 reaches the plateau. This would enable the EDL muscle as a whole to have a plateau of peak tension over a greater range of muscle length change.

There is evidence demonstrating that sarcomere length is not constant throughout the length of the muscle fiber [Huxley and Peachy, 1961; Gordon et al., 1966]; however, there has not been a systematic documentation of this inhomogeneity of sarcomere lengths between fibers within a belly or between fibers within a whole muscle. In this muscle preparation, sarcomere length was significantly shorter at the proximal end of the muscle than most of the distal half with the maximal difference in sarcomere length (10%) being similar to other reports [Huxley and Peachy, 1961; Gordon et al., 1966]. Sarcomere length begins to decrease toward the distal end of the fiber. These results suggest that sarcomere length measurements taken near the middle of the muscle fiber should provide a good estimate of the sarcomere length over 70% of the total fiber. The muscles were fixed at a position approximating the neutral position of the joint while the animal was under anesthesia. The sarcomere length in the middle portion of the fiber, in this neutral position (2.86 ± 0.14 μm), is similar to sarcomere lengths recorded at peak isometric tension (2.6–2.7 μm) [Crow and Kushmerick, 1983; James et al., 1995]. In this particular case, the muscle length at the neutral position of the associated joints is similar to the length at the plateau of the isometric length-tension curve.

The mouse EDL thus represents a multitendoned design in which there is architectural specialization between the distinct muscle bellies. This specialization may be a design feature that allows the whole EDL to generate peak tension over a greater muscle excursion during locomotion.

Acknowledgments

The authors wish to thank Rebecca Chamberlain for her excellent artistic work, Tom Burkholder for his assistance with the mathematical modeling, and Odile Mathieu-Costello for her generous assistance in the perfusion fixation of the animals. This work was supported by NIH grant AR40050 and the Department of Veterans Affairs.

References

Balice-Gordon, R.J., W.J. Thompson (1988) The organization and development of compartmentalized innervation in rat extensor digitorum lonus muscle. J Physiol *398:* 211–231.

Brand, P.W., R.B. Beach, D.E. Thompson (1981) Relative tension and potential excursion of muscle in the forearm and hand. J Hand Surg *3:* 209–219.

Brooks, S.V., J.A. Faulkner (1988) Contractile properties of skeletal muscles from young, adult and aged mice. J Physiol *404:* 71–82.

Burkholder, T.J., B. Fingado, S. Baron, R.L. Lieber (1994) Relationship between muscle fiber types and sizes and muscle architectural properties in the mouse hindlimb. J Morphol *221:* 177–190.

Cavagna, G.A., M. Kaneko (1977) Mechanical work and efficiency in level walking and running. J Physiol *268:* 467–481.

Close, R.I. (1972) Dynamic properties of mammalian skeletal muscles. Physiol Rev *52:* 129–197.

Crow, M.T., M.J. Kushmerick (1983) Correlated reduction of velocity of shortening and the rate of energy utilization in mouse fast-twitch muscle during a continuous tetanus. J Gen Physiol *82:* 703–720.

Gans, C. (1982) Fiber architecture and muscle function. Exerc Sport Sci Rev *10:* 160–207.

Gordon, A.M., A.F. Huxley, F.J. Julian (1966) The variation in isometric tension with sarcomere length in vertebrate muscle fibers. J Physiol *184:* 170–192.

Hoffer, J.A., A.A. Caputi, I.E. Pose, R.I. Griffiths (1989) Roles of muscle activity and load on the relationship between muscle spindle length and whole muscle length in the freely walking cat. Progr Brain Res *80:* 75–85.

Huxley, A.F., L.D. Peachy (1961) The maximum length for contraction in vertebrate striated muscle. J Physiol *156:* 150–165.

James, R.S., J.D. Altringham, D.F. Goldspink (1995) The mechanical properties of fast and slow skeletal muscle of the mouse in relation to their locomotory function. J Exp Biol *198:* 491–502.

Katz, B. (1939) The relation between force and speed in muscular contraction. J Physiol *96:* 45–64.

Lieber, R.L., Y. Yeh, R.J. Baskin (1984) Sarcomere length determination using laser diffraction. Biophys J *45:* 1007–1016.

Lieber, R.L., M.D. Jacobson, B.M. Fazeli, R.A. Abrams, M.J. Botte (1992) Architecture of selected muscle of the arm and forearm: Anatomy and implication for tendon transfer. J Hand Surg *17:* 787–798.

Luff, A.R. (1981) Dynamic properties of the inferior rectus, extensor digitorum longus, diaphragm and soleus muscles of the mouse. J Physiol *313:* 161–171.

Lundberg, S.R., S.T. Enguidanos, S.K. Kramer, R.A. Abrams, G.J. Loren, R.L. Lieber (1997) Tendon biomechanical properties correlate with human digital flexor muscle function. J Biomech, submitted.

Mathieu-Costello, O. (1987) Capillary tortuosity and degree of contraction or extension of skeletal muscles. Microvasc Res *33:* 98–117.

Mendez, J., A. Keys (1960) Density and composition of mammalian muscle. Metabolism *9:* 184–188.

Morgan, D.L., U. Proske, D. Warren (1978) Measurements of muscle stiffness and the mechanism of elastic storage of energy in hopping kangaroos. J Physiol *282:* 253–261.

Nicolopoulos-Stournaras, S., J.F. Iles (1984) Hindlimb muscle activity during locomotion in the rat. J Zool *203:* 427–440.

Pearse, A.G.E. (1961) Histochemistry: Theoretical and Applied, ed 4, vol 2. Edinburgh, Churchill Livingstone.

Peter, J.B., J. Barnard, V.R. Edgerton (1972) Metabolic profiles of three fiber types of skeletal muscle in guinea pigs and rabbits. Biochemistry *11:* 2627–2633.

Powell, P.L., R.R. Roy, P. Kanim, M. Bello, V.R. Edgerton (1984) Predictability of skeletal muscle tension from architectural determination in guinea pig hind limbs. J Appl Physiol *57:* 1715–1721.
Sacks, R.D., R.R. Roy (1982) Architecture of hind limb muscle of cats: Functional significance. J Morphol *173:* 185–195.
Serlin, D.M., M.H. Schieber (1993) Morphologic regions of the multitendoned extrinsic finger muscle in the monkey forearm. Acta Anat *146:* 255–266.
Warren, G.L., D.A. Hayes, D.A. Lowe, J.H. Williams, R.B. Armstrong (1994) Eccentric contraction-induced injury in normal and hind-limb-suspended mouse soleus and EDL muscles. J Appl Physiol *77:* 1421–1430.
Wickiewicz, T.L., R.R. Roy, P.L. Powell, V.R. Edgerton (1983) Muscle architecture of the human lower limb. Clin Orthop Rel Res *179:* 275–283.
Zuurbier, C.J., P.A. Huijing (1992) Influence of muscle geometry on shortening speed of fibre, aponeurosis and muscle. J Biomech *25:* 1017–1026.

Announcement

Symposium Internationale Tertium Anatomiae Clinicae

Under the patronage of the European Association of Clinical Anatomy

Medical University of Varna,
Department of Anatomy, Histology and Embryology
Bulgarian Anatomical Society

Varna, October 9–11, 1998

General Information

We cordially invite you to attend to the Symposium Internationale Tertium Anatomiae Clinicae, which will be held from October 9 to 11, 1998, in Varna.

Topics

Papers and posters from all branches of clinical anatomy, including clinical anthropology, diagnostic immunohistochemistry and immunocytochemistry will be accepted.

The official languages of the symposium will be English, German and French.

Registration fee

Active participants	DM 80.–
Accompanying persons	DM 40.–
Students	Admission free

The registration fees comprise the following:
(1) Opening ceremony and an informal get-together in the evening
(2) Congress material
(3) Coffee
(4) Dinner (10 October, 1998, 19.30 h)

Instructions to Authors

The deadline for the submission of abstracts is May 31, 1998

Abstracts must be typed in English, German or French. The name of the presenting author should appear first. Each abstract should include a brief introduction outlining the aim of the work, a short description of the experimental procedures and a summary of the data and conclusions.

Please mail your abstract together with two copies to:

Prof. Dr. sc. med. Spassimir Nikolov
Department of Anatomy, Histology and Embryology
Medical University of Varna
55, Marin Drinov Street
BG–9002 Varna (Bulgaria)
Tel. +359 52 221206, Fax +359 52 222584
E-Mail anatomia@asclep.muvar.acad.bg

All abstracts will be reviewed, but only those authors whose abstracts are not considered suitable for inclusion will be notified.

Accepted abstracts will be published in a supplement to *Scripta Scientifica Medica* before the meeting.

Acta Anatomica

Acta Anat 1997;159:156–166

A. Ishihara[a]
A. Hori[b]
R.R. Roy[d]
Y. Oishi[e]
R.J. Talmadge[d, e]
Y. Ohira[c]
S. Kobayashi[b]
V.R. Edgerton[d, e]

Perineal Muscles and Their Innervation: Metabolic and Functional Significance of the Motor Unit

[a] Laboratory of Neurochemistry, Faculty of Integrated Human Studies, and
[b] Department of Neuroscience, Graduate School of Human and Environmental Studies, Kyoto University, Kyoto,
[c] Department of Physiology and Biomechanics, National Institute of Fitness and Sports, Kanoya, Japan;
[d] Brain Research Institute, School of Medicine, and
[e] Department of Physiological Science, University of California, Los Angeles, Calif., USA

Key Words
Levator ani
Bulbocavernosus
Spinal motoneuron
Histochemistry
SDH activity
Immunohistochemistry
Rat

Abstract
Cross-sectional areas and succinate dehydrogenase (SDH) activities of muscle fibers in the rat levator ani (LA) and bulbocavernosus (BC) were determined and compared with those of the soleus (SOL) and superficial (TAs) and deep (TAd) portions of the tibialis anterior (TA). In addition, cell body sizes and SDH activities of spinal motoneurons innervating the LA and BC were examined. Histochemical myofibrillar adenosine triphosphatase (mATPase) staining reactions following alkaline and acid preincubations revealed that all the muscle fibers in the LA and BC were type IIB. Gel electrophoresis, however, showed that the LA and BC contained 2.9 and 2.4% type IIx myosin heavy chain (MHC) isoform, respectively. Immunohistochemical analyses using MHC antibodies showed that the muscle fibers in the LA and BC had types IIx+IIa (~3%) or type IIb MHC isoforms. The mean fiber cross-sectional areas in the LA and BC were significantly smaller than those in the SOL, TAs, or TAd. The mean fiber SDH activities in the LA and BC were significantly lower than those in the SOL or TAd, and similar to TAs. The population of alpha motoneurons innervating the LA and BC had similar SDH activities, irrespective of their cell body sizes. These data indicate that the LA and BC are comprised of a relatively homogeneous population of small, fast and low oxidative fibers innervated by a relatively homogeneous population of spinal motoneurons. These characteristics of the muscle fibers and motoneurons are consistent with their function in short, high-intensity activities.

KARGER

Fax +41 61 306 12 34
E-Mail karger@karger.ch
www.karger.com

0001–5180/97/1593–0156$15.00/0

This article is also accessible online at:
http://BioMedNet.com/karger

Akihiko Ishihara, PhD
Kyoto University, Laboratory of Neurochemistry, Faculty of Integrated Human Studies
Sakyo-ku, Kyoto 606-01 (Japan)
Tel. +81-75-753-6881, Fax +81-75-753-6771
E-Mail ishihara@life.h.kyoto-u.ac.jp

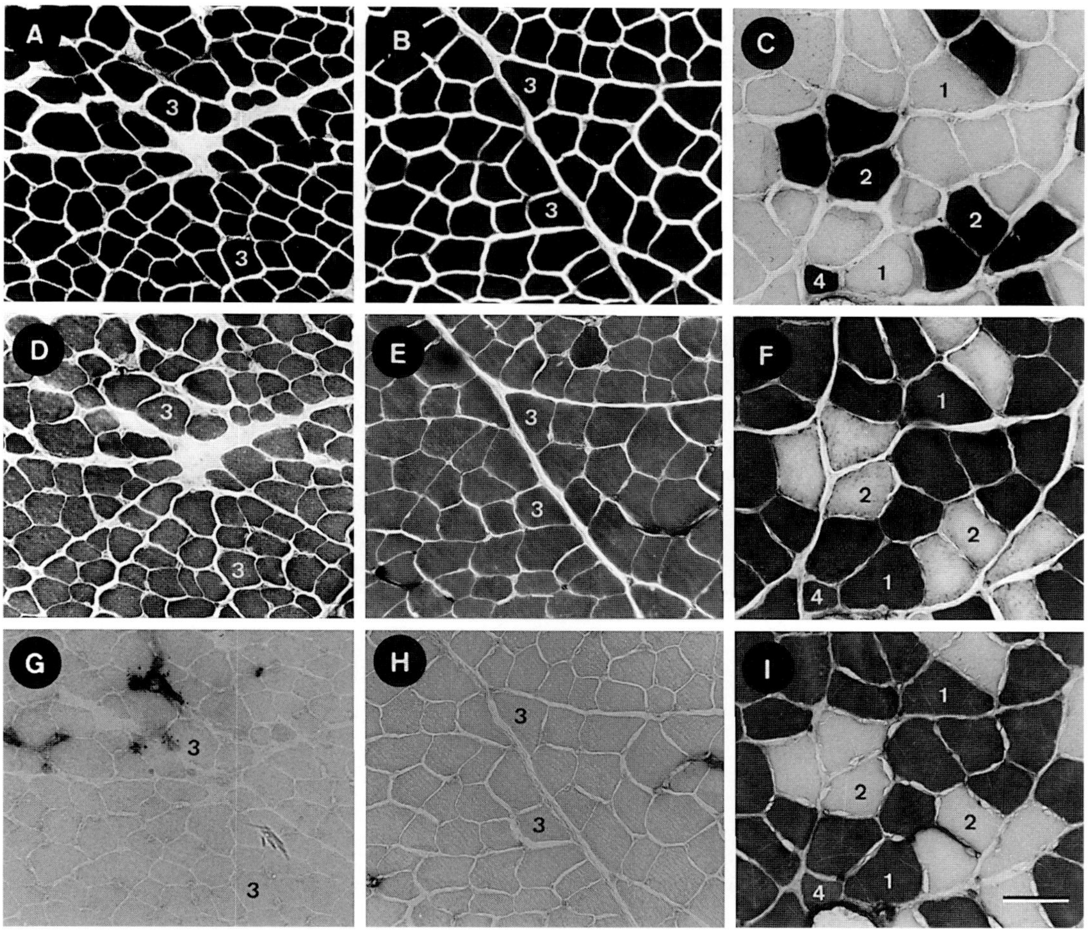

Fig. 1. Serial transverse sections of the rat levator ani (**A**, **D**, and **G**), bulbocavernosus (**B**, **E**, and **H**), and soleus (**C**, **F**, and **I**) muscles. The sections are stained for adenosine triphosphatase activity following preincubations at pHs 10.3 (**A**, **B**, and **C**), 4. 5 (**D**, **E**, and **F**), and 4.3 (**G**, **H**, and **I**). 1, type I; 2, type IIA; 3, type IIB; 4, type IIC. Scale bar = 50 μm.

Introduction

The rat perineal striated muscle complex is comprised of three muscles: the levator ani (LA), the bulbocavernosus (BC), and the ischiocavernosus. Both the LA and BC are innervated by motoneurons in the bulbocavernosus nucleus located in the ventromedial region of the spinal cord [Breedlove, 1984]. The LA attaches exclusively to the base of the penis and encircles the rectum and the BC wraps around the base of the penis and induces an erection of the glans penis with a flared cup [Breedlove, 1984]. The muscle fibers are designed to contract rapidly for short periods. There is little information related to the oxidative enzyme activities of individual fibers in these two muscles. Recently, Blanco et al. [1995] reported that all the muscle fibers in the LA and BC of male rats were histochemically type IIB, based on the staining patterns of muscle fibers for myofibrillar adenosine triphosphatase (mATPase) following alkaline (fast fibers stain darkly and slow fibers stain lightly) and acid (fast fibers stain lightly and slow fibers stain darkly) preincubations. In addition, they observed that muscle fibers in the LA and BC showed relatively low succinate dehydrogenase (SDH) activities compared to those in the SOL and the deep portion of the medial gastrocnemius. The predominance of type IIB fibers in these perineal muscles is consistent with their normal activity patterns. For example, these muscles are recruited for relatively brief periods at high activation levels during intromission and ejaculation [Holmes et al., 1991].

Using gel electrophoresis, Talmadge and Roy [1993] showed that the rat LA was comprised of 98% type IIb and 2% type IIx myosin heavy chain (MHC) isoforms. Muscle fibers containing type IIx MHC isoform are reported to have higher SDH activities than those contain-

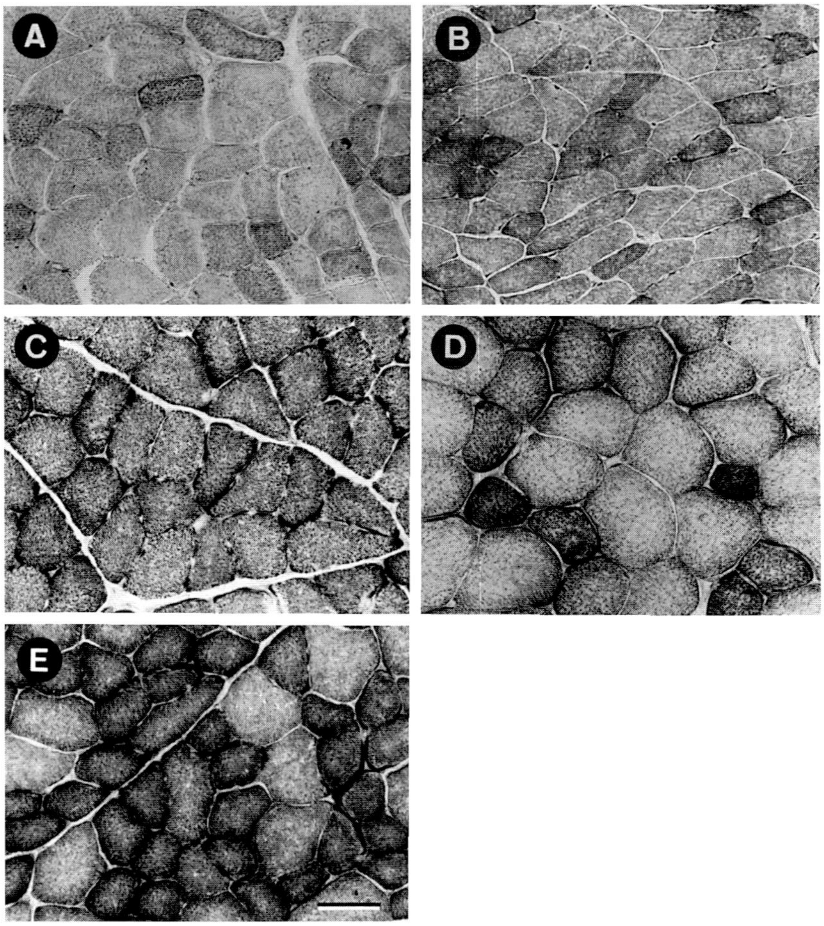

Fig. 2. Transverse sections of the rat levator ani (**A**), bulbocavernosus (**B**), soleus (**C**), and the superficial (**D**) and deep (**E**) portions of tibialis anterior muscles. The sections are stained for succinate dehydrogenase activity. Scale bar = 50 μm.

ing type IIb MHC isoform in a variety of muscles [Reichmann and Pette, 1982; Delp and Duan, 1996]. Therefore, it would be expected that muscle fibers in the LA and BC containing type IIx MHC isoform would have relatively high SDH activities. However, to our knowledge, the immunohistochemical identification of the MHC isoforms in the LA and BC muscle fibers has not been investigated. Furthermore, cell body size or SDH activity data are not available for the spinal motoneurons innervating the LA and BC. These parameters also would be useful in assessing the functional significance of motor units in these muscles.

The primary purpose of this study, therefore, was to determine the fiber cross-sectional areas and SDH activities in the LA and BC of adult male rats. These results were compared to similar data from the predominantly slow SOL and from the predominantly fast superficial (TAs, away from the bone) and mixed fast and slow deep (TAd, close to the bone) portions of the TA of the same rats. In addition, cell body sizes and SDH activities of spinal motoneurons innervating the LA and BC were examined. The results indicate that the LA and BC are comprised of relatively small and low oxidative fast fibers: the ranges in both fiber size and SDH activity were about one-half that observed in the SOL and TA. In the LA and BC, the muscle fibers containing types IIx+IIa MHC isoforms had relatively high SDH activities. The absence of gamma motoneurons innervating the LA and BC corresponds with the observation of no muscle spindles in those muscles. The alpha motoneurons innervating the LA and BC had similar cell body sizes and SDH activities.

Materials and Methods

Experimental Animals and Application of the Fluorescent Dye

Four 8-week-old Wistar male rats were used in this study. For injection of the fluorescent dye (nuclear yellow), each rat was anesthetized with sodium pentobarbital (35 mg/kg i.p.). Ten microliters of

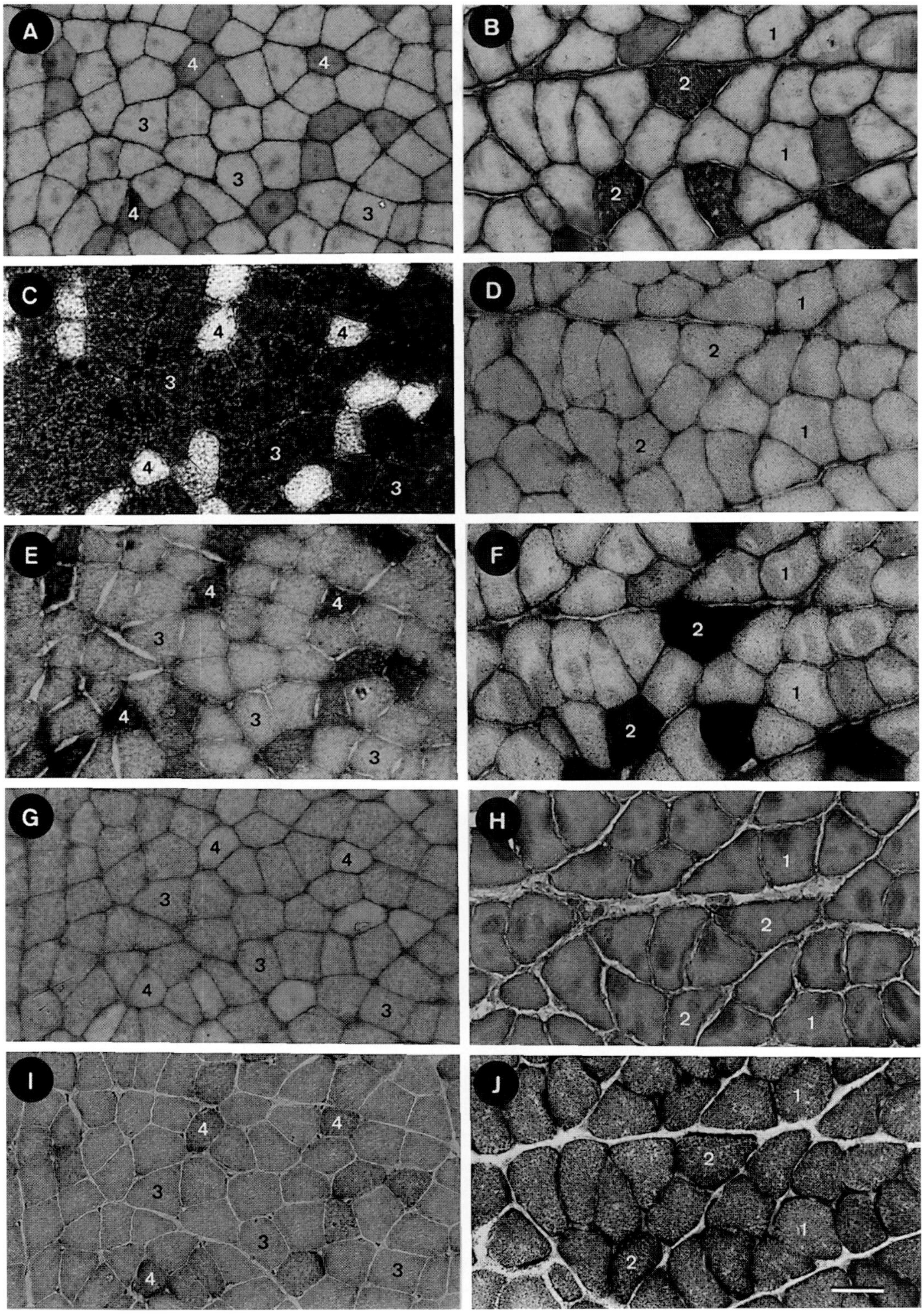

Fig. 3. Serial transverse sections of the rat bulbocavernosus (left side) and soleus (right side) muscles. The sections are reacted for type IIa (**A** and **B**), type IIb (**C** and **D**), type IIa+type IIx (**E** and **F**), and all types except type IIx (**G** and **H**) MHC isoforms, and succinate dehydrogenase (**I** and **J**). Some muscle fibers with types IIx+IIa MHC (No. 4) are present on the transverse section of the muscle. 1, type I; 2, type IIa; 3, type IIb; 4, types IIx+IIa. Scale bar=50 μm.

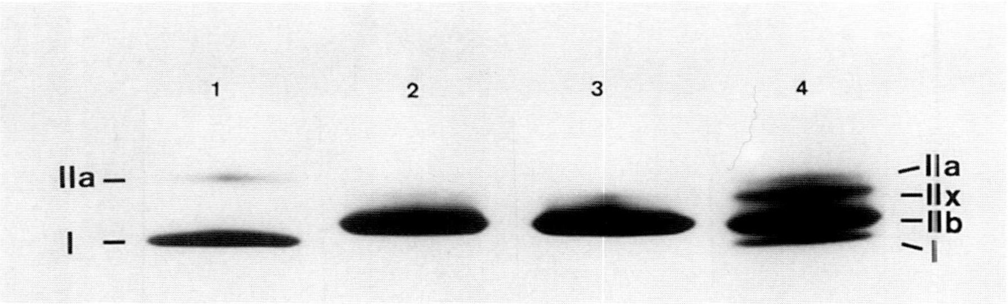

Fig. 4. Sodium dodecyl sulfate-polyacrylamide gel electrophoretic analysis of MHC isoforms in the rat levator ani (lane 2) and bulbocavernosus (lane 3) muscles. Only the region of the gels corresponding to the position of the MHC is shown. The MHC bands were labeled types IIa, IIx, IIb, and I in order of increasing electrophoretic mobility. Soleus marker (lane 1) includes types IIa and I, and the gastrocnemius marker (lane 4) includes types IIa, IIx, IIb, and I.

4

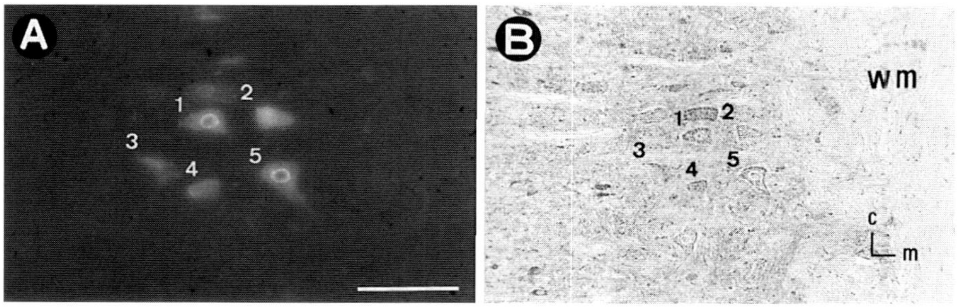

Fig. 5. Serial longitudinal sections of the rat spinal cord. Five bulbocavernosus motoneurons on the section are labeled by nuclear yellow (**A**). After identification of the motoneurons, the section was stained for succinate dehydrogenase activity (**B**). The labeled motoneurons are clustered at the ventromedial region of the ventral horn of the spinal cord. Three motoneurons (No. 2, 3, and 4) have no nucleus on the section. wm = White matter; c = caudal direction; m = medial direction. Scale bar = 100 µm.

5

a 2% solution of nuclear yellow were injected into the left LA and the same amount of a 2% solution of nuclear yellow was injected into the right SOL of 2 rats under aseptic conditions. Ten microliters and 20 µl of a 2% solution of nuclear yellow were injected into the left BC and the right TA of 2 other rats, respectively. Care was taken to ensure that the surrounding muscles were not exposed to the dye by isolating the target muscle, injecting the dye slowly to prevent reflux from the injection site and using cotton tip applicators to absorb any dye that leaked out of the muscle. One day after injection, the rats were deeply anesthetized with sodium pentobarbital (50 mg/kg i.p.). The LA, BC, SOL, and TA from both sides, and the lumbosacral enlargement of the spinal cord were removed. The muscles from the non-injected side were cleaned of excess fat and connective tissue and wet weighed. The muscles from the injected side were placed on cork, stretched to approximate their in vivo length and quickly frozen in isopentane cooled in liquid nitrogen. The lumbosacral enlargement of the spinal cord was quickly frozen in isopentane cooled in liquid nitrogen.

Enzyme Histochemistry

Serial transverse sections, 10 µm thick, from the mid-belly of the muscle were cut in a cryostat at –20 °C. Sections were brought to room temperature, air-dried for 30 min and stained for mATPase following alkaline (pH 10.3) and acid (pHs 4.5 and 4.3) preincubations [Brooke and Kaiser, 1970a, b]. The muscle fibers were classified as type I (positive at pHs 4.5 and 4.3, and negative at pH 10.3), type IIA (positive at pH 10.3, and negative at pHs 4.5 and 4.3), type IIB (positive at pHs 10.3 and 4.5, and negative at pH 4.3), and type IIC (positive at pHs 10.3, 4.5, and 4.3) (fig. 1). The muscle fiber type composition was determined from ~200 fibers on the transverse sections of the muscle.

Serial sections were stained for SDH activity, an indicator of mitochondrial oxidative potential, for 10 min at room temperature [Chalmers and Edgerton, 1989a, b] (fig. 2). The SDH activity was determined in an incubation medium containing 100 m*M* phosphate buffer (pH 7.5), 0.9 m*M* sodium azide, 0.9 m*M* 1-methoxyphenazine methylsulphate, 1.5 m*M* nitroblue tetrazolium, 5.6 m*M* EDTA-disodium salt, and 48 m*M* succinate disodium salt. The reaction was stopped by multiple washings in distilled water and the sections were dehydrated in graded alcohols, passed through xylene and then coverslipped. Histochemical control sections, in which either the succinate disodium salt or the nitroblue tetrazolium was excluded from the incubation medium, showed no positive SDH staining.

The cross-sectional area and SDH activity of ~50 muscle fibers from each muscle or muscle portion were examined using a computer-assisted image processing system (NeuroImaging System). The sections were digitized as gray scale pictures and the value of the SDH

Table 1. Fiber type compositions (%) of the rat levator ani, bulbocavernosus, soleus, and the superficial and deep portions of the tibialis anterior muscles

	n	I	IIA	IIB	IIC
LA	4	0	0	100	0
BC	4	0	0	100	0
SOL	4	86.6±3.4	11.8±3.1	0	1.7±0.6
TAs	4	0	23.4±4.6	76.6±4.6	0
TAd	4	3.0±1.6	57.1±4.9	39.9±4.8	0

Values are means ± SD. n = Number of animals; LA = levator ani; BC = bulbocavernosus; SOL = soleus; TAs = superficial portion of tibialis anterior; TAd = deep portion of tibialis anterior.

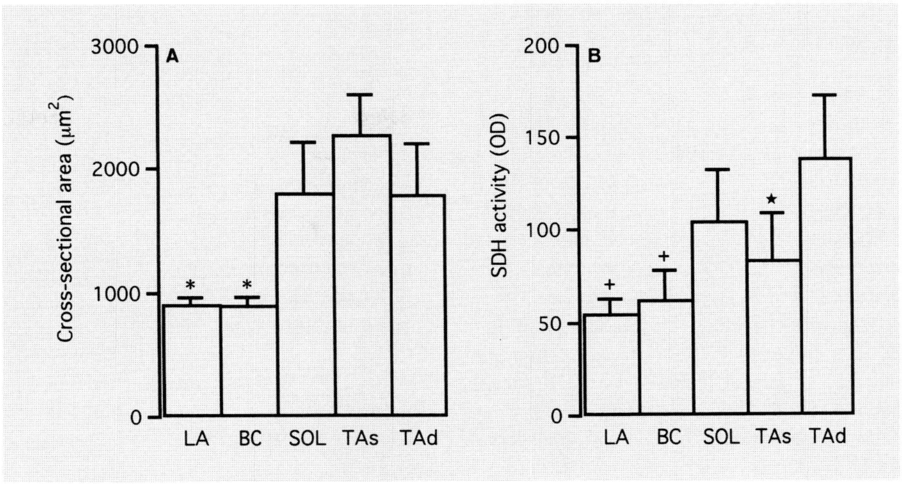

Fig. 6. Mean cross-sectional areas (**A**) and succinate dehydrogenase (SDH) activities (**B**) of the rat levator ani (LA), bulbocavernosus (BC), soleus (SOL), and the superficial (TAs) and deep (TAd) portions of tibialis anterior muscles. OD = Optical density. Values are expressed as mean and SD. * Significantly different from SOL, TAs, and TAd; + significantly different from SOL and TAd; ★ significantly different from TAd.

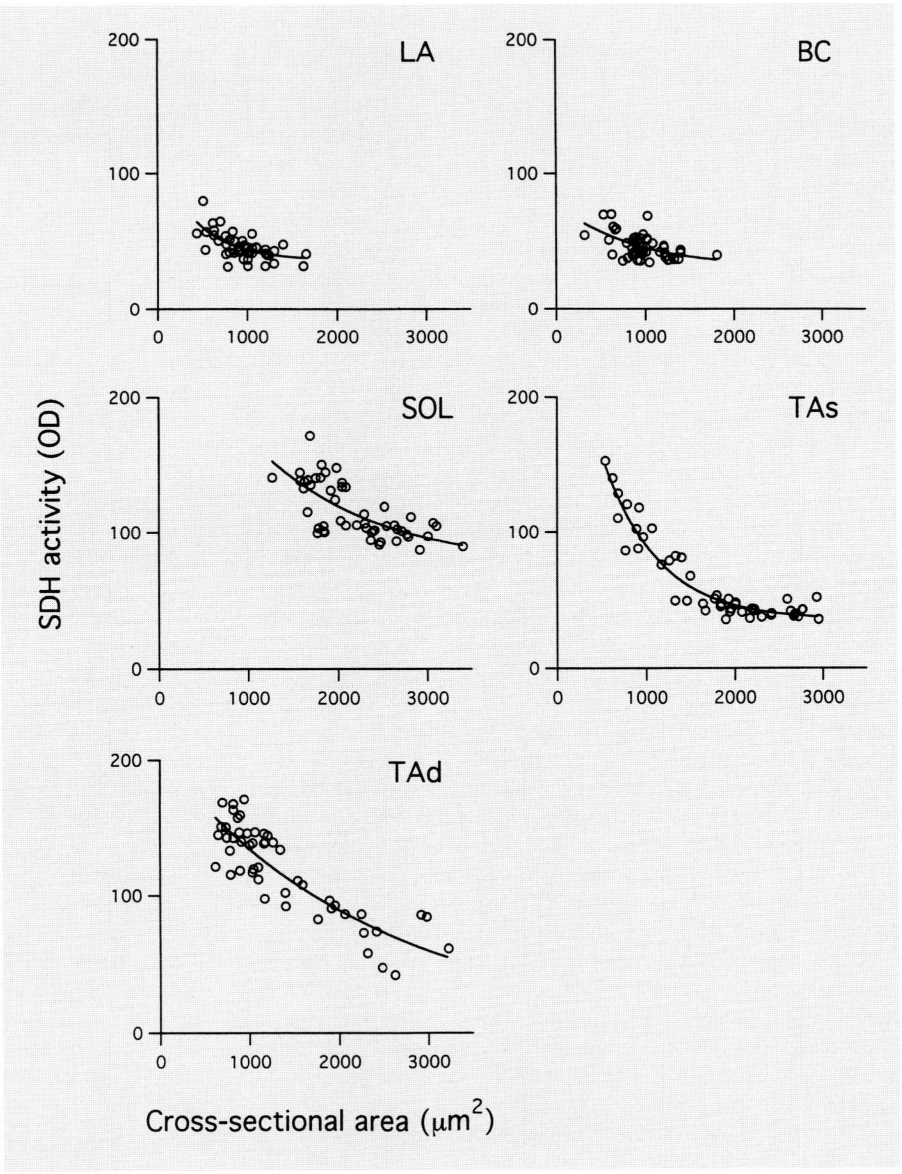

Fig. 7. Relationships betweeen cross-sectional area and succinate dehydrogenase (SDH) activity of muscle fibers in the levator ani (LA), bulbocavernosus (BC), soleus (SOL), and the superficial (TAs) and deep (TAd) portions of the tibialis anterior muscles obtained from 1 rat (No. 1). See table 2 for correlation coefficients.

staining intensity was expressed as an optical density (OD) value [Ishihara et al., 1995].

Immunohistochemistry

Immunohistochemical analysis was performed on serial transverse sections, 10 μm thick, using monoclonal antibodies specific to MHC isoforms type IIa (71), type IIb (F3), type IIa+type IIx (N1.551), and all types except type IIx (35) [Talmadge et al., 1995, 1996] (fig. 3). The sections were incubated with an optimal dilution of the primary antibody overnight at 4 °C. The avidin-biotin immunohistochemical procedure was used for the localization of MHC specific primary antibody binding according to the instructions for a PK-6102 kit (Vector Laboratories, Burlingame, Calif., USA) for IgG (antibodies 71 and 35) or an AK-5010 kit for IgM (antibodies F3 and N1.551). In control experiments, the primary antibody was omitted. The sections were incubated with the secondary antibody for 60 min for both IgG and IgM at room temperature. After incubation with the antibody, the sections were dehydrated in graded alcohols, passed through xylene and then coverslipped. Phosphate-buffered saline was used as buffer for the IgG class primary antibodies and Tris (hydroxymethyl) aminomethane-buffered saline was used as buffer for the IgM class primary antibodies.

Gel Electrophoresis

The muscles from the non-injected side were used to determine the MHC isoform composition using sodium dodecyl sulfate-polyacrylamide gel electrophoresis (SDS-PAGE) according to the method described by Sugiura and Murakami [1990] with a slight modification [Oishi, 1993; Oishi et al., 1994]. The muscle was homogenized using a glass homogenizer in 40 vol of the homogenizing buffer containing 5 *M* urea, 2 *M* thiourea, 0.13% 2-mercaptoethanol, and 10 m*M* sodium diphosphate decahydrate. The homogenate was mixed with 74 vol of SDS-sample buffer containing 250 m*M* Tris (hydroxymethyl) aminomethane (TRIS)-HCl (pH 6.8), 2% SDS, 10% glycerol, 5% 2-mercaptoethanol, and 0.02% bromophenol blue. The same sample volume (20 μl) was subjected to SDS-PAGE in a separating gel containing a linear gradient of 5–8% acrylamide and 30–40% glycerol for 24 h at 4 °C. The stacking gel contained 3.5% acrylamide and 35% glycerol. Thereafter, the gel was silver stained and the amounts of the MHC isoforms were estimated by densitometry. The MHC isoforms in the muscle were classified as types I, IIb, IIx, and IIa according to their molecular weights (fig. 4).

Spinal Motoneuron Analysis

Serial longitudinal sections, 20 μm thick, of the spinal cord were cut on a cryostat at –20 °C. The motoneurons were identified using a fluorescent microscope at 360 nm wavelength. On the untreated fresh sections, the motoneurons innervating the muscles were identified by a golden-yellow fluorescence of the nucleus with nuclear yellow (fig. 5A). The same sections used to identify the motoneurons then were stained for SDH activity, an oxidative marker enzyme, as described previously [Chalmers and Edgerton, 1989a, b] (fig. 5B).

The cross-sectional areas of the labeled motoneurons in which the nucleus was clearly visible were measured from the SDH-stained sections using a computer-assisted image processing system (Neuro-Imaging System). The sections were digitized as gray scale pictures and the value of the SDH staining intensity was expressed as an OD value [Ishihara et al., 1995].

Motoneurons with a cell body size greater than 491 μm² (i.e. 25 μm in diameter) were presumed to be alpha motoneurons innervating extrafusal muscle fibers. Smaller motoneurons were presumed to be gamma motoneurons innervating intrafusal muscle fibers within muscle spindles. The cut-off for the distinction between gamma and alpha motoneurons was primarily based on convention [Sickles and Oblak, 1984; Swett et al., 1986; Ishihara et al., 1988] and there is some imprecision using this convention. However, there is a high probability that the smaller motoneurons are gamma and the larger motoneurons are alpha.

Statistics

Means, standard deviations, and correlations were calculated from individual values using standard procedures. Values are expressed as mean ± SD. An ANOVA was used to determine overall significant differences and then a post hoc test (Scheffé's F) using StatView to determine individual differences. Significances of the regression were determined by simple and polynomial regression analyses using StatView. A 0.05 level of probability was established for statistical significance.

Table 2. Relationships between cross-sectional area and succinate dehydrogenase activity of muscle fibers in the rat levator ani, bulbocavernosus, soleus, and the superficial and deep portions of tibialis anterior muscles

Rat No.	LA	BC	SOL	TAs	TAd
1	0.466	0.601	0.683	0.957	0.854
2	NS	0.560	NS	0.651	0.880
3	0.655	0.543	0.647	0.899	0.941
4	NS	0.689	0.541	0.897	0.930

Values are correlation coefficients. NS = Not significant; LA = levator ani; BC = bulbocavernosus; SOL = soleus; TAs = superficial portion of tibialis anterior; TAd = deep portion of tibialis anterior.

Table 3. Mean cell body sizes and succinate dehydrogenase (SDH) activities of motoneurons innervating the rat levator ani, bulbocavernosus, soleus, and tibialis anterior muscles

Rat No.	n	Cell body size	SDH activity	r
1, LA_1	77	982 ± 189	0.677 ± 0.052	–0.042
2, LA_2	66	973 ± 205	0.672 ± 0.044	0.022
3, BC_1	82	1,048 ± 191	0.664 ± 0.050	–0.102
4, BC_2	86	1,063 ± 206	0.658 ± 0.045	–0.149
1, SOL_1	48	783 ± 343	0.623 ± 0.086	–0.541*
2, SOL_2	56	778 ± 328	0.633 ± 0.082	–0.527*
3, TA_1	87	868 ± 354	0.591 ± 0.083	–0.529*
4, TA_2	97	891 ± 372	0.585 ± 0.084	–0.596*

Values are means ± SD. n = Number of nuclear yellow-labeled motoneurons analyzed; r = correlation coefficient; LA = levator ani; BC = bulbocavernosus; SOL = soleus; TA = tibialis anterior.
* $p<0.05$.

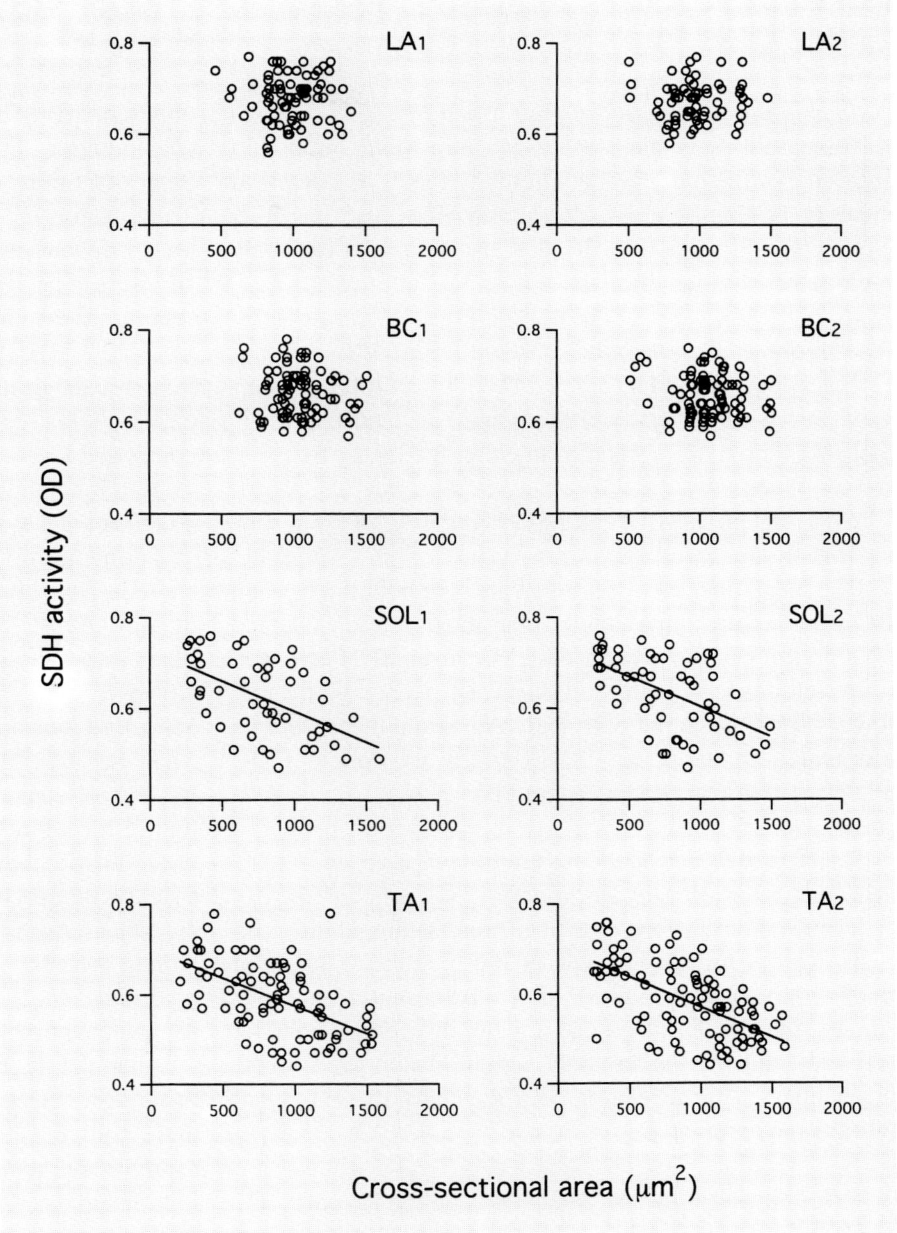

Fig. 8. Relationships between cross-sectional area and succinate dehydrogenase (SDH) activity of motoneurons innervating the levator ani (LA_1 and LA_2), bulbocavernosus (BC_1 and BC_2), soleus (SOL_1 and SOL_2), and tibialis anterior (TA_1 and TA_2) muscles. See table 3 for correlation coefficients.

Results

Body and Muscle Weights

The mean body weight of the rats was 234±3 g (n=4). The absolute weights of the non-injected LA, BC, SOL, and TA were 65±14, 77±30, 98±27, and 399±42 mg, respectively.

Enzyme Histochemical Fiber Types

All muscle fibers in both the LA and BC were classified as type IIB based on qualitative histochemical analysis, i.e. all the fibers stained positive for mATPase following preincubation at pHs 10.3 and 4.5, and negative for mATPase following preincubation at pH 4.3 (fig. 1). On the other hand, the SOL had types I, IIA, and IIC, the TAs had types IIA and IIB, and TAd had types I, IIA, and IIB (table 1).

MHC Isoforms of the LA and BC

Gel electrophoresis showed that the LA was comprised of 97.1±0.5% type IIb and 2.9±0.5% type IIx MHC isoforms. The MHC isoform composition of the BC was very similar, i.e. 97.6±0.7% type IIb and 2.4±0.7% type IIx MHC isoforms.

Immunohistochemistry of the LA and BC

Immunohistochemical analyses using MHC antibodies showed that the muscle fibers in the LA and BC contained either types IIx+IIa (~3%) or type IIb MHC isoform. Muscle fibers with types IIx+IIa MHC isoforms had relatively high SDH activities (fig. 3).

Absence of Muscle Spindles in the LA and BC

Based on serial SDH-stained sections at 100-μm intervals (every tenth section) throughout the entire muscle, there were no muscle spindles in either the LA or BC.

Cross-Sectional Areas and SDH Activities of Muscle Fibers

The mean fiber cross-sectional areas in both the LA and BC were significantly smaller than those in the SOL and both portions of the TA (fig. 6A). The mean SDH activities of muscle fibers in both the LA and BC were significantly lower than those in the SOL and TAd, while there were no differences in the mean SDH activity of muscle fibers among the LA, BC, and TAs (fig. 6B).

Relationships between Cross-Sectional Area and SDH Activity of Muscle Fibers

The relationship between fiber cross-sectional area and SDH activity for each muscle or muscle portion for 1 rat (No. 1) is shown in figure 7. For all the muscle portions studied, the smaller fibers had a wider distribution in SDH activities and the higher SDH activities compared to the larger fibers. The correlation coefficients for the individual rats are shown in table 2.

Cell Body Sizes and SDH Activities of the LA and BC Motoneurons

The motoneurons innervating the target muscles were identified by their staining for nuclear yellow (fig. 5A). Once identified, serial sections were stained for SDH activity, and cell body size and SDH activity determined (fig. 5B). Mean cell body sizes and SDH activities of the motoneurons in the motoneuron pool are shown in table 3. The SDH activities for the motoneuron pools of the LA and BC of each muscle were similar, irrespective of cell body size (fig. 8).

Discussion

In this study, we found that both the LA and BC had only type IIB fibers based on routine mATPase histochemical staining reactions and that the muscle fibers had lower SDH activities compared to fibers in the SOL and TAd. These results are consistent with a recent study [Blanco et al., 1995] that reported that all muscle fibers in the LA and BC were type IIB based on their histochemical mATPase staining intensities and those muscle fibers showed relatively low SDH activities compared to muscle fibers in the SOL and the deep portion of the medial gastrocnemius. Gel electrophoresis analyses, however, showed that type IIx comprised 2–3% of the MHC in the LA and BC in this study. This finding is consistent with the previous study of Talmadge and Roy [1993].

Furthermore, using immunohistochemical procedures, we found that all the fibers having type IIx MHC isoform in the LA and BC also contained type IIa (fig. 3A) and these fibers invariably had relatively high SDH activities (fig. 3I). However, these hybrid fibers (containing types IIx+IIa MHC) are considered to have a small amount of type IIa MHC because this isoform was not detected in the LA and BC by gel electrophoresis and those hybrid fibers stained lightly for antibody 35 (fig. 3G). The high SDH activities of the hybrid fibers containing types IIx+IIa MHC are consistent with previous studies on a variety of hindlimb muscles which report that SDH activities of muscle fibers containing type IIx are intermediate to those of type IIa and type IIb fibers [Larsson et al., 1991; DeNardi et al., 1993; Hämäläinen and Pette, 1993; Delp and Duan, 1996].

Although the muscle fibers in the LA and BC had relatively low SDH activities, the motoneurons innervating these muscles had relatively high SDH activities. For example, the mean SDH activities of the alpha motoneurons innervating TAd and TAs have been reported to be 0.601±0.012 OD and 0.526±0.023 OD, respectively [Ishihara et al., 1995], whereas the mean SDH activities of the alpha motoneurons were 0.677 OD and 0.672 OD for the LA of 2 rats and 0.664 OD and 0.658 OD for the BC of 2 rats in the present study. In addition, the SDH activities were more homogeneous in the LA and BC motoneurons than in the SOL and TA motoneurons in this study (fig. 8) and in the previous studies using the SOL and TA motoneurons [Ishihara et al., 1988, 1990, 1991, 1994, 1995]. For example, previous studies have shown an inverse relationship between motoneuronal soma size and SDH activity in the motoneuron pools of the SOL [Ishihara et al., 1988, 1990] and TA in rats [Ishihara et al., 1991], and in the deep and superficial portions of the TA in rats [Ishihara et al., 1994,

1995]. Furthermore, motoneurons innervating slow fibers had higher mean SDH activities than those innervating fast fibers across or within a muscle, suggesting an interdependence in the SDH activity between the motoneurons and their target muscle units within a motoneuron pool.

In this study, the SDH activities of the motoneurons and muscle fibers associated with the LA and BC were mismatched: the SDH activities were relatively high in the motoneurons whereas they were relatively low in the muscle fibers. This observation raises the issues of which functions of the LA and BC motoneurons might impose a greater metabolic demand and if there is a mismatch between the energy needs of these neurons and their metabolic potential. In this light it is interesting to note that there appears to be at least a 4-fold range in the estimated size of motor units in the LA [Jordan et al., 1992], but no differences in the cell body sizes or SDH properties of the innervating motoneurons [present study].

The mean cross-sectional areas of the fibers in the LA and BC were small compared to the mean fiber size across fiber types in the hindlimb muscles studied. This finding was somewhat surprising since type IIB fibers usually have the largest cross-sectional areas within a muscle comprised of a mixture of fiber types. In contrast, the range in soma sizes of the motoneurons innervating the LA and BC was similar to that reported for alpha motoneurons innervating hindlimb muscles comprised of a mixture of fiber types and sizes as shown in this study (fig. 8). Furthermore, motoneurons with a cell body size less than ~500 μm^2 were not observed in the LA and BC motoneuron pools. These smaller motoneurons are routinely presumed to be gamma motoneurons innervating intrafusal muscle fibers. The absence of gamma motoneurons in the LA and BC corresponds with our present observation of no muscle spindles in these muscles. The absence of muscle spindles indicates that the muscle contractions occur without proprioceptive regulation of muscle length and/or force via Ia pathways.

Acknowledgements

We thank Dr. S. Schiaffino (University of Padova, Italy) for providing us with the monoclonal antibodies 71, F3, and 35. The monoclonal antibody Nl.551 developed by Dr. H. Blau was obtained from the Developmental Studies Hybridoma Bank maintained by the Department of Pharmacology and Molecular Sciences, the Johns Hopkins University School of Medicine and the Department of Biology, University of Iowa. This work was supported, in part, by Grant-in-Aids for Scientific Reseach (07780074, 07680111, and 08680112) from the Ministry of Education, Science, and Culture, Japan.

References

Blanco, C.E., P.E. Micevych, W.-Z. Zhan, G.C. Sieck (1995) Succinate dehydrogenase activity of sexually dimorphic muscles of rats. J Appl Physiol *78:* 2147–2152.

Breedlove, S.M. (1984) Steroid influences on the development and function of a neuromuscular system. Prog Brain Res *61:* 147–170.

Brooke, M.H., K.K. Kaiser (1970a) Muscle fiber types: How many and what kind? Arch Neurol *23:* 369–379.

Brooke, M.H., K.K. Kaiser (1970b) Three 'myosin adenosine triphosphatase' systems: The nature of their pH lability and sulfhydryl dependence. J Histochem Cytochem *18:* 670–672.

Chalmers, G.R., V.R. Edgerton (1989a) Marked and variable inhibition by chemical fixation of cytochrome oxidase and succinate dehydrogenase in single motoneurons. J Histochem Cytochem *37:* 899–901.

Chalmers, G.R., V.R. Edgerton (1989b) Single motoneuron succinate dehydrogenase activity. J Histochem Cytochem *37:* 1107–1114.

Delp, M.D., C. Duan (1996) Composition and size of type I, IIA, IID/X, and IIB fibers and citrate synthase activity of rat muscle. J Appl Physiol *80:* 261–270.

DeNardi, C., S. Ausoni, P. Moretti, L. Gorza, M. Velleca, M. Buckingham, S. Schiaffino (1993) Type 2X-myosin heavy chain is coded by a muscle fiber type-specific and developmentally regulated gene. J Cell Biol *123:* 823–835.

Hämäläinen, N., D. Pette (1993) The histochemical profiles of fast fiber types IIB, IID, and IIA in skeletal muscle of mouse, rat, and rabbit. J Histochem Cytochem *41:* 733–743.

Holmes, G.M., W.D. Chapple, R.E. Leipheimer, B.D. Sachs (1991) Electromyographic analysis of male rat perineal muscles during copulation and reflexive erections. Physiol Behav *49:* 1235–1246.

Ishihara, A., H. Naitoh, H. Araki, Y. Nishihira (1988) Soma size and oxidative enzyme activity of motoneurones supplying the fast twitch and slow twitch muscles in the rat. Brain Res *446:* 195–198.

Ishihara, A., R.R. Roy, V.R. Edgerton (1995) Succinate dehydrogenase activity and soma size of motoneurons innervating different portions of the rat tibialis anterior. Neuroscience *68:* 813–822.

Ishihara, A., S. Taguchi, H. Araki, Y. Nishihira (1991) Oxidative and glycolytic metabolism of the tibialis anterior motoneurons in the rat. Acta Physiol Scand *141:* 129–130.

Ishihara, A., S. Taguchi, M. Itoh, K. Itoh (1990) Oxidative metabolism of the rat soleus neuron pool following hypobaric hypoxia. Brain Res Bull *24:* 143–146.

Ishihara, A., H. Tsuzimoto, H. Suzuki, N. Kasuga (1994) Postnatal changes in cell body size and oxidative enzyme activity of spinal motoneurons innervating the rat tibialis anterior muscle. Dev Brain Res *83:* 28–34.

Jordan, C.L., P.A. Pawson, A.A. Arnold, A.D. Grinnell (1992) Hormonal regulation of motor unit size and synaptic strength during synapse elimination in the rat levator ani muscle. J Neurosci *12:* 4447–4459.

Larsson, L., T. Amsved, L. Edström, L. Gorza, S. Schiaffino (1991) Effects of age on physiological, immunohistochemical and biochemical properties of fast-twitch single motor units in the rat. J Physiol *443:* 257–275.

Oishi, Y. (1993) Relationship between myosin heavy chain IId isoform and fibre types in soleus muscle of the rat after hindlimb suspension. Eur J Appl Physiol *66:* 451–454.

Oishi, Y., H. Yamamoto, E. Miyamoto (1994) Changes in fibre type composition and myosin heavy chain IId isoform in rat soleus muscle during recovery period after hindlimb suspension. Eur J Appl Physiol *68:* 102–106.

Reichmann, H., D. Pette (1982) A comparative microphotometric study of succinate dehydrogenase activity levels in type I, IIA and IIB fibres of mammalian and human muscles. Histochemistry *74:* 27–41.

Sickles, D.W., T.G. Oblak (1984) Metabolic variation among α-motoneurons innervating different muscle-fiber types. I. Oxidative enzyme activity. J Neurophysiol *51:* 529–537.

Sugiura, T., N. Murakami (1990) Separation of myosin heavy chain isoforms in rat skeletal muscles by gradient sodium dodecyl sulfate-polyacrylamide gel electrophoresis. Biomed Res *11:* 87–91.

Swett, J.E., R.P. Wikholm, R.H.I. Blanks, A.L. Swett, L.C. Conley (1986) Motoneurons of the rat sciatic nerve. Exp Neurol *93:* 227–252.

Talmadge, R.J., E. Grossman, R.R. Roy (1996) Myosin heavy chain composition of adult feline *(Felis catus)* limb and diaphragm muscles. J Exp Zool *275:* 413–420.

Talmadge, R.J., R.R. Roy (1993) Electrophoretic separation of rat skeletal muscle myosin heavy-chain isoforms. J Appl Physiol *75:* 2337–2340.

Talmadge, R.J., R.R. Roy, V.R. Edgerton (1995) Prominence of myosin heavy chain hybrid fibers in soleus muscle of spinal cord-transected rats. J Appl Physiol *78:* 1256–1265.

Acta Anatomica

List of Reviewers Vol. 159, No. 2–3, 1997

Botterman, B.R., Dallas, Tex., USA

Caiozzo, V.J., Irvine, Calif., USA

Grossman, E.J., Los Angeles, Calif., USA

Hermanson, J.W., Ithaca, N.Y., USA
Herring, S.W., Seattle, Wash., USA

Ishihara, A., Kyoto, Japan

Lieber, R.L., San Diego, Calif., USA

Monti, R.J., Los Angeles, Calif., USA

Pierotti, D.J., Flagstaff, Ariz., USA

Roy, R.R., Los Angeles, Calif., USA

Trotter, J., Albuquerque, N.Mex., USA

Author Index Vol. 159, No. 2–3, 1997

Acta Anatomica

Subject Index Vol. 159, No. 2–3, 1997

KARGER

Fax + 41 61 306 12 34
E-Mail karger@karger.ch
www.karger.com